Digital S.O.B.E.R. Method

Beat Phone Addiction

and Reclaim Your Attention in 5 Stoic Steps

Digital S.O.B.E.R. Method

Beat Phone Addiction
and Reclaim Your Attention in 5 Stoic Steps

Massimo Rigotti

Rigotti Press
Lincoln, Nebraska
United States of America

Disclaimer: The information provided by the author in this book is for informational purposes only and is not intended as medical advice, diagnosis, or treatment. The author is not a medical professional and the information contained herein should not be used to diagnose or treat any health or medical condition. Any decision you make related to your health and medical treatment should be made in consultation with a qualified healthcare professional. The author and publisher expressly disclaims any liability, loss, or risk, personal or otherwise, which may be incurred as a result, directly or indirectly, of the use or application of any of the contents of this material. Always consult with a healthcare provider before making any decisions related to your health. The S.O.B.E.R. Method™ and other processes in this book are not an endorsement and may not work for the reader. For more information about Flavors of Confidence S.O.B.E.R. Method™ and tools, please visit sobermethod.com.

Published by Rigotti Press
Lincoln, Nebraska, United States
www.rigottipress.com

Digital S.O.B.E.R. Method: Beat Phone Addiction and Reclaim Your Attention in 5 Stoic Steps

Distributed by Rigotti Press.

For ordering information or special discounts for bulk purchases, please contact via email at info@rigottipress.com or Rigotti Press, P.O. Box 6745, Lincoln, NE 68506-0745, United States of America.

Cover design by Massimo Rigotti and Elise Anderson.
Cover photograph by Brea Coan.

Library of Congress Control Number: 2026916827

Paperback ISBN-13: 979-8-9875533-6-7
Hardback ISBN-13: 979-8-9875533-7-4
eBook ISBN-13: 979-8-9875533-8-1

dedicated to

Samantha Alison Thomas

1995 – 2020

...whose foresight about these little glowing rectangles,
which we first recognized together in 2014,
proved more accurate than either of us
could have admitted at the time.

Contents

Preface

Here is a truth I did not expect to be writing.

When *Flavors of Confidence: S.O.B.E.R. Method* was published, I imagined its readership would be people fighting the same battles I fought: alcohol, drugs, the profound shame and self-disintegration that comes with substance addiction. I was delighted to be wrong. Letters began arriving from people who had never touched a drug harder than a glass of wine, people who described themselves as "normal" by any conventional measure, yet who were losing something precious: their presence, their attention, their ability to sit quietly with themselves without immediately reaching for a phone.

One letter in particular stopped me cold. A reader, and I will not share identifying details, wrote to say that they had worked the S.O.B.E.R. Method™ to break what they called their "device addiction." They described years of compulsive scrolling, the phantom vibrations they felt even when their phone was across the room, the slow erosion of meaningful relationships because no dinner table, no conversation, no moment of genuine quiet was ever allowed to remain undisrupted. They described picking up their phone to check the time and putting it down forty-five minutes later without knowing where those forty-five minutes had gone. They described the shame. The self-contempt. The sincere, repeated attempts to stop that somehow never stuck.

They described, in other words, something that felt exactly like addiction.

I reread that letter three times. Then I went and found my notes from conversations with Samantha Alison Thomas back in 2014. I had been writing about those conversations for years, and this time I was reading them in a new light. Because Samantha and I had, in fact, discussed this. We had predicted it, with a clarity that now feels a little eerie. We deduced that smartphones combined with social media had introduced something genuinely new into the human experience: a constant, pocket-sized mechanism for social comparison, interruption, and dopamine reward that had no precedent in the evolutionary history of the human brain. We agreed it would quietly damage confidence, particularly in young people. We agreed it would corrode the ability to be genuinely present. We did not know then what the research would eventually confirm: that by the mid-2020s, roughly 57% of American adults would self-report being addicted to their phones,[1] that meta-analyses covering hundreds of thousands of participants would link problematic smartphone use to elevated anxiety, depression, and sleep disruption,[2,3] and that a researcher named Jonathan Haidt would write a book called *The Anxious Generation*[4] documenting what many of us already suspected in our bones.

We were right. I wish we had been wrong.

This book exists because the S.O.B.E.R. Method works, and because device addiction, for all its apparent differences from substance addiction, operates through many of the same behavioral and neurological loops. The triggers are different. The drug of choice is a glowing rectangle rather than a bottle. There is no physical detox risk. Though, interestingly, two participants in my early studies exhibited what appeared to be genuine withdrawal shakes when separated from their devices for extended periods. There is no legal system to navigate. But the essential problem is the same: a habituated behavior that has migrated from conscious choice into subconscious automation, progressively displacing the life you actually want to be living.

The five-step S.O.B.E.R. Method cycle, which encompasses Stoic, Observe, Behavior, Execute, and Restore, applies with remarkable precision to this challenge. The Deming-cycle philosophy of continuous, measured improvement applies perfectly. The emphasis on a physical journal, on radical honesty with yourself, on behavior chain analysis, on building a network of people who hold you accountable. All of it translates.

What did not translate without modification were the specific examples, the specific chains of behavior, and the specific interventions. The device addict does not need guidance on how to detox from alcohol. They need guidance on phantom vibration syndrome, on notification architecture, on the psychological machinery of infinite scroll, on why their phone feels like a missing limb when it is in another room, and on what to actually do with their hands and their attention during the unstructured minutes that used to be filled by checking Instagram, TikTok, or whatever your app of choice happens to be.

That guidance is what this book provides.

I want to be honest with you about what this book is not. It is not a lecture about technology being bad. Smartphones are extraordinary tools. They have connected families across continents, saved lives through emergency services, given voice to people who previously had none, and made a staggering amount of human knowledge available to anyone with a data connection. The problem is not the device. The problem is the relationship. And the relationship, for tens of millions of people, has quietly become one of compulsion rather than choice.

If you picked up this book, something in you already knows. That knowing is enough. Let's work with it.

The S.O.B.E.R. Method is not a quick fix, and I will not pretend otherwise. It is a continuous-improvement cycle you will run from now until it becomes so fundamental to who you are that you begin applying it on the fly, in real time, in moments you never thought you'd be able to navigate with a phone in your pocket and not in your hand. The first time that happens, the first time you sit through a quiet moment, or a difficult conversation, or a long wait, without reaching for the device, you will know something has shifted. Something real. Something that belongs to you.

That is your Flavor of Confidence. Let's go find it.

Massimo Rigotti

Method Overview

The Flavors of Confidence S.O.B.E.R. Method™ is the most unique process available for achieving and maintaining what I am calling digital sobriety: a healthy, intentional, self-directed relationship with technology, in which your devices work for you rather than the other way around.

Let me be clear about what digital sobriety means in this context, because clarity matters here. It does not mean throwing your phone in a river. It does not mean living off-grid or foreswearing all social media forever. It means reaching a place where you make genuine, conscious choices about when, why, and how you engage with your devices, rather than operating on a behavioral autopilot that was programmed by app designers, notification architects, and algorithm engineers who are paid to maximize your engagement, not your well-being.

That distinction, between chosen engagement and compelled engagement, is everything.

The S.O.B.E.R. Method™ functions as a continuous self-improvement program. Unlike programs where you complete one step and move to the next, S.O.B.E.R. works like The Deming Cycle. Dr. W. Edwards Deming, the American engineer and statistician sent to postwar Japan to revolutionize its manufacturing sector, developed the concept of continuous improvement as a cycle of plan, do, check, and act; it is a loop you run indefinitely, getting better with each pass. His ideas helped transform Japan into an industrial powerhouse. I believe the same philosophy, applied inward, can transform a human life.

When I invented this method, sitting on the floor of my bedroom on the last night of 2016, holding Samantha's self-portrait in my lap, I asked myself a question that would not leave me alone: *Why can't I observe a character challenge within myself, plan action to correct it, measure whether I had, and then seek forgiveness from those directly impacted by that flaw in the past?* That question became the S.O.B.E.R. Method.

Flavors of Confidence S.O.B.E.R. Method™

The five steps are:

S: Stoic: Ground yourself in Stoic philosophy. Identify what is within your control and what is not. Apply the four cardinal virtues, wisdom, courage, justice, and temperance, to your digital life. Begin your SOBER Journal.

O: Observe: Track your actual device behavior with rigorous honesty. Use Screen Time, Digital Wellbeing, and your journal to understand your patterns: the triggers, the vulnerable moments, the behaviors you do not even realize you are doing.

B: Behavior: Design concrete changes. Modify your notification architecture, your routines, your physical environment. Build behavior chains that route you toward intentional engagement rather than compulsive reaction.

E: Execute: Implement your plan. Measure results against your goals. Adjust based on data. Journal daily. Stay accountable.

R: Restore: Forgive yourself for time and presence you surrendered. Seek forgiveness from those in your life whose attention you owe. Mentor others. Volunteer. Return to S and begin the cycle again.

This book is designed to be compact enough to carry with you. Use it as your reference guide and as a living tool. The blank pages at the end of each section are there for a reason.

First: get a physical journal. I call it your SOBER Journal throughout this book. I have kept daily journals for decades in my Ampad 22-157 notebooks. Use whatever you love: a Moleskine, a legal pad, a hardback with a cover that makes you want to pick it up. Just make it physical.

People ask me constantly whether they can type notes on their computer or use a journaling app on their phone. My answer is always the same: no. And in this particular adaptation of the method, the irony of using your phone as your primary journal while trying to develop a healthier relationship with your phone should speak for itself. The act of writing by hand, putting a pen or pencil to paper and guiding it across the page, triggers deeper contemplation. It engages your body in the cognitive process. It slows you down enough to actually think. Research on learning consistently shows that students who handwrite notes retain material more deeply than those who type them.[5] Your SOBER Journal is not a productivity tool. It is a thinking tool. Keep it away from your devices. Let it be analog in a digital world.

Second: be honest with yourself. This is non-negotiable. Your life experience is unique. Your relationship with your devices is unique. The goal of the Observe step is to see yourself clearly, not to see the version of yourself you wish you were. The people I have mentored who struggle most with this method are not those who have the deepest device habits; they are those who cannot bring themselves to tell the truth about those habits on the page.

Third: this is going to be difficult. You will want to quit. You will slide back into old patterns. You will pick up your phone without thinking and not realize what you have done until twenty minutes have passed. That is part of the process. The method is continuous improvement, not perfection. You start again. Every time.

I have mentored people using the S.O.B.E.R. Method for goals entirely unrelated to substance addiction: improving professional confidence, deepening relationships, managing anxiety, restructuring their daily lives. The results have consistently surprised me. People find greater confidence, more rewarding relationships, better sleep, a sharpened ability to be present in their own lives, and, frequently, less digital device use, even when that was not the primary goal. The method works because it addresses the root. Device addiction, like substance addiction, is almost never really about the device. It is about what the device is providing that the person cannot yet access another way: relief from boredom, escape from anxiety, the chemical warmth of social validation, the illusion of connectedness, a break from the difficulty of being alone with one's own thoughts.

When you develop the capacity to meet those needs in healthier ways, the compulsion to reach for the phone loses its grip. Not all at once. Gradually. With practice.

That is the S.O.B.E.R. difference.

Now, let's build your unique Flavor of Confidence in the digital world.

What are Flavors of Confidence?

No one is quite sure yet they sound delicious!

Credit for the term belongs to the late Samantha Alison Thomas. It is the title of a self-portrait she gave me in 2014, in which she described herself as "one who feigned confidence, just made of portions of a fake whole." It is my most treasured possession. Tragically, Samantha was murdered by a drunk driver on August 1, 2020, in Mobile, Alabama. I think about her every day, and I suspect she would have had a great deal to say about the topic of this book; none of it polite, most of it exactly right.

Back in 2014, Samantha and I spent hours, sometimes the better part of a day, on FaceTime calls we called our "mind explorations." We were, among other things, trying to figure out what confidence actually was. We started with the dictionary. I still use my paperback Webster's from 1984. It's a little worn with a couple spine breaks yet it functions as perfectly now as it did in 5th grade.

> **con•fi•dence** (kän'fə dəns) *n* **1.** trust; reliance **2.** assurance **3.** belief in one's own abilities **4.** belief that another will keep a secret

We are talking about entry 3. And even the scholars felt a little uncertain writing that one, did they not? Confidence is nearly impossible to define except you know it when you feel it exude from someone standing in front of you.

Samantha suggested we approach it from the opposite direction: figure out what confidence is not. She said immediately, "Fear!" And then we were off.

Here is what I want to tell you about fear, anxiety, and the device problem, because it connects to what Samantha and I worked out all those years ago, and it connects with startling accuracy to what the research now shows.

We deduced that anxiety had become the dominant emotional weather of modern life, particularly among young people. Samantha called out FOMO, the fear of missing out, as one of its core manifestations. And we traced FOMO's sudden intensification to exactly one thing: the smartphone combined with social media. Prior to iPhones and the social feed, you might have heard secondhand that you missed a party. You brushed it off. By 2014, when we were having these conversations, you could watch the party in real time, from inside a different party, on a device that never left your hand.

There is FOMO again. I want you to consider what FOMO actually is beneath the surface. It is the real-time conviction that whatever you are currently doing or experiencing is less worthy than what someone else is doing or experiencing. It is, in other words, a direct attack on your confidence. You cannot feel FOMO about a situation and simultaneously feel secure in your own value and choices. They are mutually exclusive states.

Now consider that many people check their phones 80 to 150 times per day.[6] Consider that the average American adult now spends between four and seven hours per day on their phone, not counting computer-based screen time.[7] Consider that every one of those checks is a potential FOMO trigger, a potential social comparison, a potential algorithm-served image designed by professionals to make you feel inadequate, envious, or left behind.

You are not accidentally feeling less confident. You are being systematically exposed to a confidence-eroding experience dozens of times every hour, and you are carrying the mechanism of that exposure in your pocket.

Samantha and I were sitting with this problem in 2014, without knowing the scale it would eventually reach. She was on her way to becoming a physician. She wanted to understand the brain chemistry. I had been diving into Stoicism and found in Marcus Aurelius a framework that seemed to address exactly the kind of internal mastery we were looking for.

"I believe that if we could run a standard test to set a baseline across a broad spectrum of a patient's chemical makeup," she told me once, "a drug could be developed specifically for that person." I pushed back, saying it was too expensive and too inaccessible, and we landed somewhere more practical: could we achieve remarkable improvement through diet, exercise, and deliberate philosophical practice? We thought we could. Years of experience with the S.O.B.E.R. Method have confirmed it.

But here is what I want to name clearly for the purposes of this book, because it has particular force in the context of device addiction.

A device habit that has become compulsive is not a confidence problem; it is a confidence generator. The phone provides a steady supply of something that feels like confidence: social validation, information, stimulation, the sense of being connected, the illusion that you are staying current with the world. The problem is that this supply is entirely external and entirely unreliable. A like can appear and then disappear. A post can receive fewer responses than you hoped. An algorithm can show you someone who seems to be living the life you want. And each of these micro-events registers in your nervous system as a small threat.

Over time, people who rely on external validation from devices for their sense of confidence find themselves in a deepening paradox: the more they check, the more they need to check; the more they post, the more vulnerable they feel to the response. The very tool they are using to feel better is making them feel worse, and the answer their nervous system proposes, more checking, more scrolling, more posting, intensifies the problem.

This is the Flavor of Confidence problem in the digital age. You have been building a Flavor of Confidence out of borrowed ingredients: other people's reactions, algorithmic approval signals, and the dopamine pulse of a notification. The work of this book is to help you build a Flavor of Confidence from ingredients that actually belong to you.

"Do you think anxiety is actually the largest cause?" Samantha asked me on a call.

"I'm not sure," I said. "You can be anxious about your appearance, whether you fit in, whether you are good enough…"

She said, "I think it is."

For device addiction specifically, I have come to agree. The compulsive phone use is, in most cases, anxiety management. People reach for their phones when they are bored, which is anxiety about unstructured time. They reach for them when they are in a social situation that is uncomfortable, which is anxiety about connection. They reach for them when they have a hard task in front of them, which is anxiety about capability. They reach for them in the middle of the night when they cannot sleep, which is anxiety about…anything, really. The device has become the universal emotional regulator, available at any hour, at zero apparent cost.

The cost, of course, is not zero. It is your attention, your presence, your sleep, your relationships, and in time your confidence, because genuine confidence is built only one way: through the repeated experience of navigating difficulty without an escape hatch. Every time you reach for the phone instead of sitting with discomfort, you teach your nervous system that the discomfort is intolerable. Every time you scroll instead of engaging with the present moment, you practice the habit of absence.

We also, in those 2014 conversations, discovered something about the relationship between confidence and having options. Having a backup plan, even a vague one, empowers decisive action. Anxiety is in large part the experience of feeling trapped with no alternatives. Here is an interesting thing the S.O.B.E.R. Method will teach you: once you begin to develop genuine capacity to manage your own inner states without the device, you discover that you have more options than you realized. The discomfort was survivable. The quiet was not actually empty. You were capable of more than your nervous system believed.

When you get there, and I am confident you will, that is your Flavor of Confidence in the digital realm. Not perfect digital minimalism. Not the absence of a smartphone. Rather: the settled, experienced knowledge that you are the one choosing how to engage. That is a confidence no algorithm can take away.

The First 48 Hours

Before we begin the real work, I want to give you something to hold.

I have already promised you that the S.O.B.E.R. Method is not a quick fix, and I meant it. The chapters ahead will ask you to ground yourself in Stoic philosophy, to observe your own behavior with a level of honesty that will sometimes sting, and to rebuild your daily life from the foundation up. That work is the point, and it takes time. There is no shortcut, and anyone who sells you one is selling you something that isn't recovery.

But I have mentored enough people to know something about the first few days. The reader who feels one small, real shift in the first 48 hours is far more likely to still be working the Method in the first 48 days. Early momentum matters. A single quiet morning, or a single dinner where you actually heard the person across from you, can be the difference between a book you finish and a book that ends up face-down on the nightstand, next to the phone you meant to leave in the kitchen.

So let me be very clear about what this chapter is, because the distinction matters. This is not the Method. These are not the deep interventions, and they will not fix anything on their own. They are a few first, reversible steps, the equivalent of opening a window in a stuffy room. What they will do is buy you a little quiet. And that quiet is exactly where the real work begins in the very next chapter.

Here is what I am asking you to do in the next two days. You do not need to fully understand why each one works yet. You will, once you have read the chapters that follow. For now, I am asking you to act, lightly, and to pay attention to what happens inside you when you do.

1. **Tonight, move your phone charger out of your bedroom**. Plug it in the kitchen, the hallway, anywhere that is not arm's reach from your pillow. If you use your phone as an alarm, this is the night you dig out an old alarm clock or buy a cheap one. I will make the full case for this later. For now, trust me: this single change is one of the highest-return actions in the entire book, and you can do it before you go to sleep tonight.

2. **Turn off every notification that is not a real human being**. Go into

your settings and silence them all: the games, the news, the shopping apps, the social feeds, the email that does not actually need you in real time. Keep calls, keep texts from people you love, keep your calendar. Everything else can wait six hours, and you will not miss it. You are not deleting anything. You are simply deciding to stop being summoned.

3. **Switch your phone to grayscale**. It takes about thirty seconds, and the instructions are in the Behavior chapter if you want them now. Stripped of its engineered color, your phone becomes a tool again instead of a slot machine. Leave it in gray for the next two days and notice how much less it pulls at you.

4. **Eat one meal, phone in another room**. Just one, sometime in the next 48 hours. If you can do it with someone you care about, even better. Notice what the conversation feels like. Notice what the food tastes like. Notice the urge to reach for a phone that is not there, and notice that the urge passes.

And one more, because it is the bridge to everything else:

5. **Get a physical journal and write one honest line in it**. Not a paragraph. One line, about why you picked up this book in the first place. That journal is the single most important tool in the Method, and the next chapter will explain why. Starting it tonight means you arrive at the real work already holding the pen.

That is the whole list. Five small things, none of which require you to understand the philosophy yet, and all of which you can undo tomorrow if you choose.

Now, a word about what to expect, because I do not want the next two days to surprise you. You may feel restless. You may reach for a phone that is in another room and feel a small jolt of something that is almost panic. You may notice how often your hand moves toward your pocket out of pure habit, with no thought attached to it at all. This is normal. It is not a sign that something is wrong with you. It is a sign that you are noticing, for perhaps the first time, just how automatic the habit has become. That restlessness is not your enemy. It is information, and in the very next chapter you are going to learn how to read it.

Do not try to be perfect in these 48 hours. You will slip. You will pick the phone up without meaning to. That is fine. There is no zero in this Method, and there is nothing to start over. When you notice you have slipped, simply set the phone back down and carry on. That is the whole skill, and you are already practicing it.

When the two days are up, you will not be cured. But you will have proof, small and real and yours, that the quiet is survivable, that the urge passes, and that you are capable of choosing. That proof is worth more than any promise I could make you on this page.

Now let's build the foundation underneath it.

Notes

Your Daily S.O.B.E.R. Practice

Everything in this book eventually comes down to what you do on an ordinary day. The chapters ahead will explain the reasoning behind each piece. This page is the piece itself: the daily rhythm of digital sobriety, gathered in one place so you never have to wonder what the practice actually asks of you.

You will not do all of it perfectly, and you are not meant to. Treat it as the shape of a good day, and aim at the shape.

Morning

» No phone for the first hour. Get up, make the bed, eat, move your body. Let your own mind wake before the world's noise arrives.

» Ten minutes in your SOBER Journal. One passage of Meditations, and a line on how it meets your day.

» One intentional phone check when the hour is up. Purposeful, not a scroll.

Through the Day

» Communication on a schedule, not on demand. Pick your check-in times and keep to them.

» Phone-free meals. Every one you can manage.

» Move your body. Five days a week, at minimum.

» Notice the reaches. Each time your hand moves toward the phone without a reason, that is a note for the journal.

Evening

» A device cutoff you hold to. The last hour before sleep is phone-free.

» Review your behavior chains. Note what worked and what pulled at you.

» One line of gratitude for something that could not be photographed or posted.

» Phone charges outside the bedroom. A book on the nightstand instead.

The non-negotiables, if you remember nothing else: no phone the first hour, no phone the last hour, no phone at the table, and the journal every day.

Notes

Stoic

Stoicism, an ancient philosophy born in the Hellenistic period (323 BC to 31 BC), offers profound insights into living a virtuous life marked by self-control, resilience, and inner peace. As you begin your journey toward digital sobriety, I encourage you to work with purpose to follow its principles.

I know what some of you are thinking: *What does a 2,000-year-old Greek philosophy have to do with putting down my phone?* Everything, as it turns out. The Stoics were deeply preoccupied with a problem that is identical to ours, just expressed in different clothes: how do you remain anchored in your values and your reason when the world around you is constantly offering easier, more immediately gratifying alternatives? The Roman emperor Marcus Aurelius wrote his *Meditations* as personal field notes in the battle against distraction, procrastination, the pull of pleasure, and the anxiety of not knowing what others thought of him. He was fighting, in other words, exactly the kind of war you are fighting with your phone, and his weapons are still sharp.

Stoicism is built on four cardinal virtues: wisdom, courage, justice, and temperance. These virtues form the foundation for ethical living and personal growth, making them indispensable in the pursuit of digital sobriety. Here is a quick look at each, and then we will move into the practical application of each as it relates to your life with technology.

Wisdom (Sophia) is the first and foremost Stoic virtue. It involves the cultivation of rational understanding, self-awareness, and insight into how the world actually works, not how we wish or fear it to work. In the context of Stoicism, wisdom encompasses the Dichotomy of Control: recognizing what is within our power to change and what is not. For device addiction, wisdom means understanding the mechanics of how apps and platforms are designed to capture your attention, and using that understanding to make informed decisions rather than reactive ones.

Courage (Andreia), the second virtue, is the quality of facing fear and adversity with inner strength. Stoicism teaches that courage is not the absence of fear but the ability to act virtuously in spite of it. For device addiction, courage means doing the hard thing repeatedly: closing the app when your thumb wants to scroll further, putting the phone in another room when every habit in your body says keep it close, and sitting with the discomfort of not knowing whether someone responded to your message. These acts are small in scale. They are not small in their cumulative effect.

Justice (Dikaiosyne), the third virtue, encompasses fairness, integrity, and ethical conduct in all interactions. For our purposes, justice includes treating yourself justly, with self-compassion rather than shame, and recognizing your obligations to those around you. When you are at a dinner table looking at your phone instead of the people who showed up to be with you, that is a question of justice. When your child is talking to you and your eyes are on a screen, that is a question of justice.

Temperance (Sophrosyne), the fourth virtue, is synonymous with self-control and moderation. Stoicism teaches that excessive attachment to any pleasure disrupts inner peace and leads to a kind of moral deterioration. Temperance does not demand abstinence from technology; it asks for balance, for the ability to govern desire and impulse rather than being governed by them. This is perhaps the most directly applicable Stoic virtue to device addiction.

As you understand and integrate these four virtues, the aim is to achieve eudaimonia (a flourishing life) and ataraxia (peace of mind). Stoics believe that by focusing on these inner virtues rather than external outcomes, likes, follower counts, and notification responses, individuals can find genuine tranquility regardless of what the algorithm gives them.

Take five minutes right now. Close your eyes. Reflect on what you just read. Think about which of the four virtues you will struggle with most in the context of your device use. Most people I work with immediately identify temperance. Some say courage. Welcome back.

The Stoic Dichotomy of Control

The Stoic Dichotomy of Control focuses on the distinction between things within our control and those outside it. Epictetus, writing in his *Enchiridion*, made the essential statement: some things are up to us, and others are not.

For device addiction, applying this principle is both clarifying and somewhat uncomfortable, because part of us would like the problem to be external, located in the phone or the apps rather than in our choices. Let me walk you through the distinction carefully.

1. **Recognize What You Can Control.**
   - **Your choices about engagement.** You can choose whether to open an app. You can choose whether to respond to a notification immediately or at a scheduled time. You can choose to leave your phone in another room during dinner. These are genuine choices, even when they do not feel like it.
   - **Your judgments about your device use.** You can choose not to measure your self-worth by your follower count. You can choose to disengage from social comparison. You can choose to interpret a friend's lack of response to your message as anything other than a statement about your value.
   - **Your physical and digital environment.** You can configure your devices to support your intentions rather than undermine them: notifications off, grayscale mode, no social apps on your home screen, phone in a drawer during family time. These structural changes are within your control, and their cumulative effect is substantial.
   - **Your attention.** Right now, in this moment, you are choosing where to direct your cognitive resources. That choice is always yours, even though devices are engineered to make it feel otherwise.
2. **Accept What You Cannot Control.**
   - **The apps themselves.** You cannot personally restructure the engagement architecture of Instagram or TikTok. You cannot prevent the algorithm from showing you content engineered to keep you scrolling. You cannot make other people stop posting. Accept this. Your energy spent raging at the platform design is energy not spent on the one thing you can actually change: your relationship to it.

- **Other people's posting habits and online behavior.** Someone you care about will post something that makes you feel envious, or hurt, or left out. This will happen. You cannot control it. You can control whether you open the app at all, and you can control how you interpret and respond to what you see.
- **Your past device use.** You cannot recover the hours you spent scrolling. You cannot undo the times you were at a dinner or a concert or a child's recital and looked at your phone instead of the moment in front of you. That time is gone. The Stoic counsel here is neither to dwell on it nor to pretend it did not happen, but to let it inform the choices you make from this moment forward.
- **The existence of the technology.** Smartphones, social media, and the attention economy are not going away. Accepting that your challenge is not to defeat technology but to develop a mature relationship with it is the beginning of real progress.

Take another five minutes. Reflect on the Dichotomy of Control in your own digital life. In your SOBER Journal, make two columns: what you can control, and what you cannot. Be honest.

The Four Cardinal Virtues Applied to Digital Life

Wisdom in Practice

Wisdom in the context of digital sobriety begins with understanding. The apps you use most compulsively are not accidental. They are the products of teams of engineers, psychologists, and behavioral economists whose explicit professional goal was to make them as engaging as possible. The variable reward schedule, the fact that you do not know whether the next scroll will bring something interesting or mundane, is borrowed directly from slot machine design. The notification badge is engineered to create a mild anxiety state that can only be resolved by checking. The infinite scroll removes the natural stopping point that a finite page would provide.

This is not a conspiracy theory. It is documented in company internal research, in the congressional testimony of former platform employees, and in the published work of researchers like B.J. Fogg, who developed persuasive technology design at Stanford.[8] You are not weak for having been captured by these systems. You are human. But wisdom now requires you to see the architecture clearly and stop cooperating with it unconsciously.

Now let us look more deeply at how the virtue of wisdom applies, point by point, to your digital sobriety.

1. **Understand the root causes of your device use.** Wisdom begins with introspection. Underneath a compulsive habit there is almost always a deeper reason you are reaching for the screen: relief from boredom, escape from a difficult emotion, the ache of social comparison, anxiety about an unstructured moment, or the simple need to feel validated or informed. With wisdom you can discern which of these is actually driving you and seek the right remedy for it, rather than treating the surface behavior. This is not a judgment of yourself. It is information, and it is exactly what you will act on in the Behavior step.
2. **Navigate the challenges with perspective.** Digital sobriety is rarely a straight line. There are stretches of real success and there are slides back into old patterns. Wisdom carries you through both, offering perspective in the valleys and humility on the peaks. Instead of reacting impulsively to a craving or a trigger, wisdom allows for a pause, an honest assessment, and a decision made in service of your long-term goals rather than the short-term pull of the moment.
3. **Seek and accept help**. The wise understand that they do not have all the answers, and that asking for help is a strength rather than a weakness. Whether it is confiding in a trusted friend, finding a mentor, or building an accountability relationship, wisdom helps you acknowledge your need for support and, just as importantly, discern the right kind of support for your particular challenge. That tailored approach is exactly why you will find the S.O.B.E.R. Method so powerful.
4. **Accept a long-term commitment**. This is not a two-week challenge. The pathways your device habits have worn into your

brain over years will not dissolve in a weekend. Wisdom means understanding and accepting that, setting realistic expectations, celebrating small victories, learning from setbacks, and measuring your progress over months rather than days.

The Stoics believed that our greatest battles are most often the ones we wage with ourselves. It is with wisdom that you find the clarity and perspective to win those battles, and to live in line with your own highest values rather than the design of an engagement algorithm.

Courage in Practice

Courage, in the context of digital sobriety, is required in ways that may surprise you.

It takes courage to put your phone in a drawer during dinner and leave it there, even when the anxiety of not knowing what you might be missing makes your hands want to reach. It takes courage to be the person at the party who is actually present rather than photographing it. It takes courage to have a difficult conversation in person rather than retreating into the safety of a text thread. It takes courage to sit with boredom, which most of us have not practiced in years, because we have been filling every moment of unstructured time with our phones.

And perhaps most of all: it takes courage to look at your Screen Time or Digital Wellbeing data honestly, to see in black and white how many hours you spent scrolling this week, and not to rationalize or minimize or immediately close the app. That data is uncomfortable. Looking at it anyway is a small act of courage with large consequences.

To the Stoics, courage was never merely physical bravery. It was the quality that let a person confront and overcome adversity, internal and external, while keeping virtue in view. It meant standing firm in your convictions and facing your fears head-on. Device addiction asks for exactly this kind of courage, and it asks for it on three fronts.

1. **Facing your inner demons.** Compulsive device use almost always masks something underneath: boredom you would rather not feel,

anxiety you would rather not sit with, loneliness, or a difficult task you are avoiding. The courage to acknowledge and work through whatever the scrolling is covering, without the crutch of the screen, is the deepest courage this work asks of you.

2. **Enduring the discomfort**. There is no physical detox here as there is with a substance, and no medical risk of the kind that requires a doctor. But the discomfort is real: the restlessness, the low-grade anxiety, the nagging sense that you are missing something, even the phantom vibration of a phone that has not buzzed. Courage equips you to endure that discomfort and to recognize it for what it is, a temporary state on the path to something far better.
3. **Navigating social pressure**. You will sit in rooms where everyone is on a phone and the unspoken expectation is that you will be too. The courage to go against that grain, to decline the easy retreat into your screen, to be the one who looks up, and to gently distance yourself from the patterns that keep pulling you back, is quiet, but it is real.

Here are a few practical ways to cultivate that courage:

1. **Self-reflection**. Regularly examine your triggers, your fears, and your motivations. This self-awareness is the foundation courage is built on, and you will do this work in depth in the Observe and Behavior steps.
2. **Education.** Understand exactly how your particular apps and platforms are built to capture you: the variable rewards, the engineered badges, the bottomless feed. Knowing what to expect prepares your mind for the challenge and strips the machinery of some of its power.
3. **Community.** Surround yourself with people who respect and encourage this journey. They lend you strength when your own courage wavers. In the S.O.B.E.R. Method, this includes seeking out a mentor, which we take up in the Observe step.
4. **Celebrate small victories.** Every phone-free dinner, every craving you sat through, every morning you did not reach for the phone is a triumph of courage. Acknowledge these milestones, and when you

reach the larger ones, mark them and bring in the people who helped you get there.

Courage, worked daily, turns the difficult path of digital sobriety into a series of challenges you are equipped to meet, one quiet choice at a time.

Justice in Practice

Justice asks you to consider the impact of your device habits on the people around you.

Research from the University of Michigan and elsewhere has documented a phenomenon sometimes called "phubbing" (phone snubbing), the act of looking at your phone during a face-to-face interaction. Studies show that phubbing, even when the phone-user is a partner or parent, is associated with lower relationship satisfaction, increased conflict, and, in the case of parent-child interactions, measurable effects on children's emotional development.[9,10] The people you love experience your phone as competition. They are often too polite or too habituated to say so. Justice requires you to take that seriously without waiting to be told.

Justice, to the Stoics, is not only fairness in our dealings but acting with integrity, understanding our place within a community, and caring for the well-being of others. It asks you to see your device habits not as a private matter but as something that ripples outward, and it has four faces in this work.

1. **Recognize your interdependence**. Your relationship with your phone has consequences that reach well beyond you. It shapes your relationships, your reliability, your presence as a parent or partner or friend, and in some cases the literal safety of others, as anyone who has glanced at a screen behind the wheel already knows. Justice begins with seeing that interconnected web honestly and admitting that your attention, or its absence, lands on other people.
2. **Accept personal responsibility**. Justice, as the Stoics understood it, is about knowing your responsibilities and acting on them. You are accountable for the attention you have given away, and for the

commitments you have made to the people who depend on your presence. Keeping a few journal entries about how your digital sobriety benefits not just you but your relationships can instill a real sense of duty, and we will measure that progress directly in the Execute step.

3. **Benefit the people around you.** Without the constant haze of compulsive use, you bring more of yourself, your attention, your patience, your skills, your compassion, to your family and your community. Your visible, intentional technology use also quietly shifts the norms around you. Children of parents who put their phones away during family time develop a different relationship with their own devices. Later, in the Restore step, you will channel this outward through service.
4. **Seek to make amends.** Part of acting justly is repairing what was harmed. For most of us that means acknowledging, to a partner, a child, or a friend, the times we were physically present but absent behind a screen, and making it right going forward. This is powerful fuel for staying the course, and we will work on it directly in the Restore step.

Seen through the lens of justice, digital sobriety stops being only about you. It becomes a way of restoring balance to your relationships and of contributing something better to the people and the community you belong to.

Temperance in Practice

Temperance is the virtue most people reach for first when they think about device problems, and it is the one most often misapplied. Temperance is not about counting minutes or restricting yourself to a set daily screen time allowance as if technology were a dessert. It is about the cultivation of genuine self-regulation: the inner capacity to recognize an impulse, evaluate it, and choose whether to act on it.

The reason people with device problems fail at restriction approaches is not usually a lack of information or a weak desire to change. It is that restriction, without addressing the underlying need the device is meeting, creates a kind of pressure that eventually breaks. The S.O.B.E.R. Method addresses the underlying need. Temperance, worked through the Method, is therefore sustainable in a way that app-timers and phone detox challenges are not.

Through the Stoic lens, temperance offers several tools for this journey.

1. **Practice self-awareness and reflection.** Temperance starts with introspection. By regularly reflecting on your behavior, you identify the patterns of excess and the specific triggers that precede them, which lets you intervene proactively rather than react helplessly. You will deepen this work in the Observe and Behavior steps.
2. **Embrace moderation, not abstinence.** The goal is not zero screen time, as if the phone were a poison to be eliminated. The goal is screen time you choose for genuine reasons, connection, information, creativity, rest, rather than screen time that simply happens to you through compulsion. Temperance is about governing the impulse, not amputating the tool.
3. **Embrace discomfort and value the long-term reward.** Stoicism teaches that discomfort is a natural part of growth, and that delayed gratification is its own form of wealth. Instead of surrendering to the immediate urge for relief, temperance asks you to bear the discomfort of not checking, understanding it as temporary, and to keep your eyes on the lasting rewards, better sleep, deeper relationships, a clearer mind, that far outweigh the fleeting pull of the feed.
4. **Establish boundaries through self-knowledge.** Part of temperance is knowing your limits and building structures around them. The Observe step will show you exactly where your device habits do the most damage, and temperance means using that knowledge to set specific, targeted boundaries, device-free zones and times, and to have a replacement behavior ready, rather than relying on generic willpower.
5. **Cultivate resilience.** Just as temperance is about balance rather than total avoidance, digital sobriety is about resilience rather than perfection. Slips will happen. With a temperate mindset they

become lessons rather than defeats, and the work is simply to return to balance and continue. This is the continuous-improvement heart of the Method.

Temperance, practiced through the S.O.B.E.R. Method, is sustainable in a way that app-timers and detox challenges are not, because it builds the inner capacity to govern the impulse rather than merely walling off the temptation.

At this point, I would not be at all surprised if your head feels close to bursting. This is a tremendous amount of material to absorb in one pass. I genuinely recommend that you go back to the beginning of this chapter and read it again, and this time take notes in your SOBER Journal as you go. Writing these ideas in your own hand will help commit the Stoic framework to memory and give your mind a sturdier foundation to build the rest of this work on.

Meditations and Your SOBER Journal

To complete your first pass through the Stoic step, I need you to get a copy of Meditations by Marcus Aurelius. It is available free of charge at Google Books if you want a soft copy, though you already know I will push you toward a physical book, because the tactile act of reading and marking a page is a genuine assist to learning. The most easily consumed translation, the original is in Greek, is Meditations: A New Translation by Gregory Hays. It is under ten dollars, reads with remarkable clarity, and a physical copy lets you underline, dog-ear, and clip your way to the passages that resonate most. Please do not continue until you have access to a copy.

You might be wondering why on earth I am making you read a nearly two-thousand-year-old book, and how it could possibly be relevant to your life with a device that fits in your pocket. Let me give you a little history, because it makes the point better than I can.

Marcus Aurelius wrote Meditations during the later years of his life, around 170 to 180 AD, while he was on military campaigns in Central Europe. It was never intended for publication, or for any reader at all. The work is a collection of personal notes, reminders, and reflections he wrote to guide himself toward virtue, wisdom, and self-mastery. They were, quite literally, his private journal entries, a personal tool for introspection and moral improvement. Sit with that for a moment, because it is the whole reason this book is in your hands: one of the most powerful people in the known world, with access to every conceivable pleasure and distraction, kept a handwritten journal to help himself resist the pull of immediate comfort in favor of what actually mattered. That is precisely the work you are doing, and the practice you are about to begin.

The text is divided into twelve books, better understood as chapters or sections than as separate volumes, covering the nature of the universe, the impermanence of life, the discipline of attention, the ease of distraction, and the difference between what feels good now and what leads to a life well lived. Given its deeply personal nature, it is a small miracle that Meditations survived at all. The writings were preserved and copied by hand for centuries, largely by Christian monks who treasured their moral content despite their pagan origin. The first printed edition appeared in the sixteenth century, its influence grew during the Renaissance revival of Classical texts, and it has since become one of the foundational works of Stoic philosophy, precisely because its themes of resilience, virtue, and introspection are timeless. He was, in other words, working on exactly the problem you are working on.

"You have power over your mind, not outside events. Realize this, and you will find strength." – Marcus Aurelius, *Meditations 8.47*

The S.O.B.E.R. Method works as a continuous-improvement program, so treat this as a practice you are beginning rather than a book you are finishing. For your first pass through the S.O.B.E.R. cycle, read Books 1 and 2. There are thirty-four passages between them. For every day thereafter, read at least one passage, they are numbered within each book, then open your SOBER Journal and write down how it speaks to your digital sobriety and how you can adapt the idea to your own journey. Keeping these notes and rereading them regularly reinforces the Stoic foundation beneath all the practical tools you are about to build.

From those first two books, pick the three passages you feel you most need to work on. Choose the ones that challenge you most, because the obstacle is the clearest path forward. Marcus says it better than I can:

"The impediment to action advances action. What stands in the way becomes the way." Marcus Aurelius, Meditations 5.20

In your SOBER Journal, write the Book and Passage you are working through, for example Meditations 2.7 if you are reflecting on the things that distract you, and then write your reflection.

"Do external things distract you? Then make time for yourself to learn something worthwhile; stop letting yourself be pulled in all directions. But make sure you guard against the other kind of confusion. People who labor all their lives but have no purpose to direct every thought and impulse toward are wasting their time – even when hard at work." – Marcus Aurelius, *Meditations 2.7*

Here is an example of my personal reflection on *Meditations 2.7*:

It can be easy for me to seek out the stimulation of my phone in moments when I do not want to face something more difficult. I reach for it not because anything there requires my attention, but because my mind wants an exit. Marcus reminds me that the distraction and the pleasure of the moment are not the purpose of my existence. My purpose requires presence and action. When I pick up my phone to escape, I am choosing the distraction over the thing I am avoiding, and the avoided thing will still be waiting, while the time I spent scrolling will be gone. The phone is not the problem. My willingness to let it serve as a way out of my own life is the problem. From this moment, I will try: task first, phone second, and only if I have genuinely earned the rest.

Your first reflections may not reach that depth, and they do not need to. The point is to build the habit of connecting this ancient wisdom to your daily life with your devices. These reflections are the seed of the Observe step that follows, and each one becomes a thread of your first observation work.

Review your Stoic entries regularly. I read portions of mine every Sunday, and after many years they are now spread across several of my notebooks. When you eventually run out of passages in Meditations, do not stop. Re-reflect on old ones, which will surprise you with new meaning, or move on to the other Stoics, Seneca, Epictetus, Musonius Rufus, Hierocles, and Chrysippus, who will deepen your understanding of yourself further still. Know thyself. On each subsequent S.O.B.E.R. cycle, select three more challenging reflections and carry them into Observe. You will not run out. Run one again two years from now and you will find smaller, finer improvements you are only then ready to tackle.

You are an amazing machine, one that can be reshaped simply by thinking. You hold the power to improve yourself from this moment until the day you expire, toward a life of purpose and your own unique Flavor of Confidence. Every insight you record here about what drives your device behavior is a thread you will follow into the next chapter.

Notes

NOTES

Observe

I am about to ask you to do one of the most difficult things: observe your own device behavior with genuine objectivity.

Above all else, I need you to be honest with yourself as you journal these observations. And I need you to do this work alone, first. Your observations about your phone habits are going to be uncomfortable. Some of them may surprise you. Some will embarrass you. If you share them prematurely, with a partner, a friend, or a family member, you may encounter reactions that shape and color what you are willing to see. Do this work solo first. The foundation has to be built without anyone else's fingerprints on it.

Your mind will resist this, and you need to know how, because the hardest part of observing your own behavior is that the instrument doing the observing, your own mind, is not neutral. It comes with a set of built-in distortions that will quietly work against you. Naming them now is how you keep them from running the show.

Cognitive biases. Your mind is full of them. Confirmation bias will lead you to notice only the evidence that fits the story you already tell about yourself, so if you believe you have your phone under control, you will conveniently overlook the forty pickups that say otherwise. The self-serving bias will let you credit your good days to your own discipline and blame your bad days on circumstance: a stressful week, a boring meeting, everyone else being on their phones. Both biases bend your view away from the truth.

Emotions. Strong feelings cloud objectivity. Shame about how much time you have lost can make you flinch away from the Screen Time number before it even registers. Pride can make you wave off a bad week as a fluke. The feeling, whatever it is, distorts the measurement, which is exactly why we anchor everything in this step to data you cannot argue with.

The inward blind spot. When you observe other people, you catch the nuances they miss, because you have an outside vantage point. Observing yourself is inward by nature, and without that outside view you will miss your own subtlest behaviors, especially the automatic ones: the reach for the phone that you do not even register as a decision.

The subconscious habit. The behaviors you repeat most become habits, and habits sink below conscious awareness. Your device checking has been repeated so many thousands of times that it now runs underneath your notice. You cannot easily observe what you no longer consciously do, which is why this step takes deliberate, almost forensic effort.

Defense mechanisms. When you encounter behavior that threatens your self-image, your psyche defends itself, usually without asking your permission. Denial, repression, and rationalization spring up to protect you from an uncomfortable truth. Those familiar lines, "I need it for work," "everyone does it," "I could stop anytime," "it's not that bad," are not facts. They are defenses, and every one of them is a door with something behind it.

Overanalysis. There is an opposite danger too. In pursuit of self-awareness you can ruminate on every single action until you tip into analysis paralysis, which raises your stress and erodes your confidence without producing any insight. Observe honestly, write it down, and move on. You are gathering data, not putting yourself on trial.

These distortions are not the enemy, and they are not a sign that something is wrong with you. They are simply the conditions you observe under, and the way through them never changes: deliberate effort, honest reflection, and the willingness to write down what you would rather not see. When a rationalization surfaces, do not argue with it. Record it. A rationalization is a trail leading straight to the behavior it is protecting. Remember the oldest instruction in philosophy, the one said to have been carved over the temple at Delphi: know thyself. The path is winding, but the self-knowledge waiting at the end of it is the whole foundation of your Flavor of Confidence.

Let me tell you something about how your brain works with device habits, because understanding this will make the Observe step much more productive.

You operate on autopilot approximately 95% of the time. This is not a character flaw; it is a fundamental feature of a brain that has learned to conserve cognitive resources by automating familiar actions. Your morning routine, your commute, the habitual way you hold your fork: all of this runs on autopilot, freeing your conscious mind for genuinely novel problems.

The problem is that device checking has become part of your autopilot. Not a small part, either. Research by Dr. David Greenfield at the Center for Internet and Technology Addiction has documented that most people reach for their phones within 15 minutes of waking and check them dozens of times throughout the day, with many of those checks occurring *without any conscious intention to check.*[11] You are not deciding to look at your phone. Your hand is just picking it up. And the habit loop of trigger, routine, and reward runs so smoothly by now that you may genuinely not remember the specific moment you decided to open Instagram.

This is why you cannot fix the habit from the outside in. App timers help. Grayscale mode helps. But the deepest change requires understanding your specific triggers, your specific vulnerable moments, and the specific need your device is meeting, because that need will not disappear when you take the phone away. It needs somewhere else to go.

The Behavior Chain Analysis (BCA)

Behavior Chain Analysis is the most powerful tool in the Observe step. I was introduced to it during my time in rehab in 2015 and recognized it immediately as the structured version of what Samantha had been guiding me toward in her letters earlier that year, asking me to write down what I was thinking and feeling in the moment before I wanted a drink.

For device addiction, BCA works identically. You are dissecting the chain of events, thoughts, emotions, and physical sensations that lead to a specific compulsive device behavior, and identifying where in that chain you can intervene.

Here is a step-by-step guide for constructing your BCA in your SOBER Journal:

1. **Target Behavior:** Identify the specific device behavior you want to understand. Be precise. Not "I use my phone too much" but "I pick up my phone during dinner when I'm with my family" or "I scroll Instagram for thirty minutes after I wake up before I get out of bed" or "I check my phone compulsively during meetings."

2. **Prompting Event:** What triggered the target behavior? Dig into specifics. Was it a notification? Boredom? A lull in conversation? A feeling of anxiety? The physical proximity of the phone on the table?
3. **Vulnerabilities:** What conditions made you more susceptible? Were you tired? Stressed? Sitting in a location where you usually use your phone? With people you do not feel comfortable with? Alone? Had you already been using your phone earlier that day and found it easier to keep going?
4. **Sequence the Actions:** Map the full chain from trigger to the compulsive behavior. For example: Phone visible on desk → notification sound → glance at phone → unlock phone to check notification → notification is from a social app → open app → scroll for 25 minutes.
5. **Thoughts and Feelings:** At each link in that chain, what were you thinking and feeling? Perhaps: Notification sound → mild curiosity/mild anxiety about what it might be → "I'll just check this one thing" → opened app → "This is interesting, just a few more" → [pattern continues without further conscious decision-making].
6. **Short-Term Consequences:** What happened immediately as a result of the behavior? Was the itch scratched? Did you feel relief, then guilt? Did you miss something happening in the room around you? Did you lose track of time?
7. **Long-Term Consequences:** What happens if this pattern continues? Cumulatively: damaged relationships, eroded attention span, reduced productivity, disrupted sleep, a growing undercurrent of dissatisfaction with your real life compared to the curated lives on your feed?
8. **Outcome:** What actually happened? Being honest: did the scrolling session make you feel better or worse?

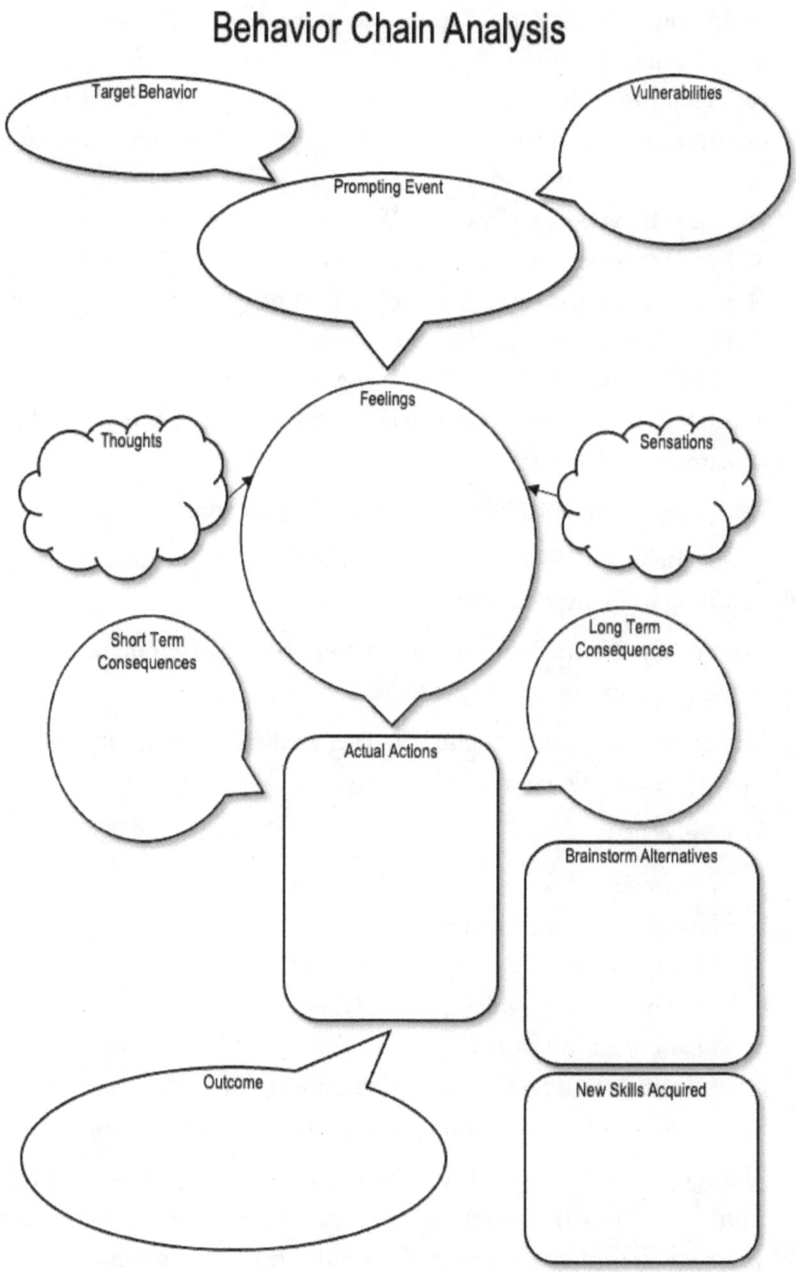
Behavior Chain Analysis
Target Behavior
Vulnerabilities
Prompting Event
Feelings
Thoughts
Sensations
Short Term Consequences
Long Term Consequences
Actual Actions
Brainstorm Alternatives
Outcome
New Skills Acquired

9. **Brainstorm Alternatives:** Given what you have mapped, what could you do differently? Not at the point of scrolling, because by then the momentum is significant, but earlier in the chain. Could you keep the phone off the desk? Could you turn off social media notifications entirely? Could you have a substitute behavior ready for the vulnerable moment?
10. **New Behaviors to Practice:** What specific, concrete new action will you try when this trigger arises? Write it down. Be specific. Not "I will try to be less on my phone" but "When I sit down to dinner, I will put my phone face-down in my bedroom before we eat."
11. **Review and Refine:** After trying the alternative, revisit your journal. Did it work? If not, where did the chain break down? What adjustment is needed?

Here is an example BCA to guide your first attempt:

Target Behavior: Picking up phone within five minutes of getting into bed, scrolling until after midnight.

Prompting Event: Lying in bed. Phone on nightstand. Body slowing down. Mind beginning to process the day.

Vulnerabilities: Tired. Slight low-level anxiety about tomorrow. Phone within arm's reach. No competing activity.

Sequence:

1. Lie down in bed.
2. Notice phone on nightstand.
3. Pick up phone "just to check the time".
4. See notifications while checking time.
5. Open one notification.
6. Continue into social feed "for just a few minutes".
7. Scroll for 45 to 90 minutes before putting phone down.

Thoughts and Feelings: "I just want to unwind." → "I'll just check this one thing." → "This is interesting, just a bit more." → time passes without awareness → "It's midnight already?" → guilt, frustrated tiredness.

Short-Term Consequences: Delayed sleep onset. Reduced sleep quality, as blue light exposure suppresses melatonin for up to 90 minutes.[12] Woke up feeling less rested than needed.

Long-Term Consequences: Cumulative sleep debt, as research links this directly to elevated cortisol, impaired decision-making, and worsened anxiety.[13] Reinforced pattern of using phone as the last activity before sleep. Nighttime device use associated with poorer mental health outcomes in multiple large-scale studies.[14]

Alternatives:

» Move phone charger to kitchen before bed.
» Replace phone with physical book on nightstand.
» Establish a 9:30 PM device cutoff as part of evening routine.

Using Built-In Tracking Tools

Before you can build a meaningful behavior chain, you need accurate data about your device use. There is no point guessing. Your built-in tools will show you the truth clearly.

iPhone Screen Time (Settings → Screen Time): Shows total daily device use, broken down by app category. Shows how many times you picked up your device and when. Shows notification volume per app. Use it weekly at minimum. The first time most people look at this honestly, it is a significant moment.

Android Digital Wellbeing (Settings → Digital Wellbeing & Parental Controls): Provides the same core data. Shows your "Unlock Count" (how many times you unlocked the device), your "Daily Total," and app-by-app breakdowns.

In your SOBER Journal, for seven consecutive days, record:

1. Total daily screen time
2. Top three apps by usage time
3. Number of pickups
4. Times of day with the highest use (Screen Time shows this)
5. Any notable trigger events that day (stress, boredom, conflict, etc.)

This seven-day baseline is not for judging yourself. It is for seeing yourself clearly. There is an enormous difference between "I think I use my phone too much" and "I picked up my phone 94 times yesterday, spent 4 hours and 12 minutes on it, and the peak usage periods were 7:15 to 8:00 AM before work and 10:30 PM to 12:15 AM in bed." One is a vague concern. The other is data you can actually work with.

Observing Your Physical and Relational State

Device addiction has physical consequences that deserve your honest inventory, just as substance addiction does.

In your SOBER Journal, write responses to these questions:

Sleep: How many hours do you sleep on average? Do you use your phone within an hour of bedtime? Do you wake up and check your phone before getting out of bed? Do you feel rested in the morning? Research consistently shows that pre-sleep device use delays sleep onset, reduces total sleep time, and degrades sleep quality, primarily through blue-light exposure(which suppresses melatonin production)[12] and cognitive arousal from social content. A 2022 meta-analysis by Scott et al. analyzing data from over 125,000 participants found significant associations between nighttime smartphone use and insomnia, daytime fatigue, and poorer overall sleep quality.[3]

Attention: How long can you read continuously without checking your device? Do you find it difficult to finish a longer article, a chapter of a book, or an extended piece of work without a device interruption? Can you sit through a conversation without feeling the pull to check your phone? Research by Gloria Mark at UC Irvine documents that after a digital interruption, it takes an average of 23 minutes to fully return to a task.[15] If you are being interrupted dozens of times per day by your phone, your brain is essentially never in a deep work state.

Anxiety: Do you feel anxious when separated from your phone? Do you experience phantom vibrations, the sensation of feeling your phone vibrate when it has not? Phantom vibration syndrome affects up to 89% of regular smartphone users in some studies, and is considered a marker of habituated phone checking.[16] Do you feel a low-grade anxiety about what you might be missing when you are not connected?

Relationships: Think about the last week. Can you recall a specific moment when you were present with someone you care about and your phone interrupted or displaced that presence? How did they respond? How did you feel afterward? Be specific.

Physical symptoms: Neck pain, headaches, eye strain, and "text neck" (forward head posture from looking down at devices) are documented physical consequences of extended device use. Do you experience any of these?

Diet: How and when do you actually eat? Do you eat while scrolling, barely tasting the meal? Do you skip meals when you are absorbed in a screen, then graze mindlessly later? Do you reach for sugar or caffeine to chase the energy your sleep and your phone habits have drained? Note your honest patterns here. We will build a better plan for this in the Behavior step, but the change starts with seeing the current reality on the page.

Write all of this in your journal. This is your baseline. You will return to it.

Role-Playing Your Device Scenarios

Having charted several behavior chains, you are ready for a tool that takes them off the page and into your body: role-playing the situations in which your device habits are most stubborn. It is one thing to map a behavior chain on paper. It is another to navigate the trigger in real time, when the pull is live and your hand is already moving. Role-playing is the dress rehearsal that bridges the two.

1. **Quick review of your BCA.** Remember what the behavior chain gave you: the sequence of events, thoughts, emotions, and decisions that leads to a specific compulsive behavior, and the points along that sequence where you can intervene. Role-playing is how you practice the intervention before you need it.
2. **Understand the power of rehearsal.** Role-playing lets you run a real-world scenario in a controlled, no-consequence setting. By

simulating the moments that usually send you to your phone, you get to practice navigating them without anything actually riding on the outcome. It is a dress rehearsal for life, and it builds genuine confidence and resilience.

3. **Set the stage.** Enlist someone you trust, a friend, a family member, or a mentor who will take this seriously, and act out a real trigger situation with them. Do not just pick scenarios off the top of your head. Play out the trigger events from the BCAs you actually charted: the dull moment at a gathering when phones start appearing around the room, the tense conversation that sends your hand to your pocket, the meeting where you habitually check notifications, the bored wait in a line. Working through all of your charted triggers, not just the convenient ones, is what makes this powerful.

4. **Apply your BCA in the moment.** As you act it out, pay close attention to what rises in you. How do you react when the trigger lands? What internal dialogue starts up? Afterward, sit down with your journal and your behavior chain and answer honestly:
   - Did I react the way I charted in my BCA, or did the old habit take over?
   - Were my anticipated triggers accurate?
   - What emotions and thoughts surfaced?
   - Were there new emotions or triggers I had not charted?
   - How did I respond, and am I satisfied with it?

Update your BCA with whatever you learned.

5. **Refine your response.** The beauty of role-playing is that you can run it again and again. With each repetition, try a different response. Practice calmly leaving the phone in your bag, naming out loud that you are stepping away from it, or redirecting to a replacement behavior you have chosen in advance. Over enough repetitions, the new response begins to feel more natural than the old one.

6. **Debrief and reflect.** After each session, talk it through with your partner in the exercise, then write it down: what worked, what surprised you, and what still needs work. Repeat the whole process until your intended response starts to arrive on its own, before the

old habit can.

This is not comfortable, and that is exactly the point. Keep at it until the alternative responses feel more natural than the habitual ones. That is the sign you are ready to carry them out of the rehearsal and into your real life.

Communication With Your People

One of the most practically important things you can do in the Observe step is communicate your intentions to the people closest to you.

This does not need to be a dramatic announcement. It needs to be honest and clear:

> *"I'm working on being more present and less on my phone. I'm going to be putting it away during meals. I might seem a little absent sometimes when I'm fighting the urge to check it; I'm not being rude, I'm just working on this. If you notice me slipping back into old habits, it's ok to say something."*

What you are doing here is a few things simultaneously. You are making a commitment out loud, which increases accountability. You are explaining your behavior so that the people around you do not misinterpret it. And you are creating an opening for them to be part of your support system rather than innocent bystanders to your private struggle.

When you make this communication, do it face-to-face if possible, not by text. No smartphones in the room. It is a useful place to start practicing what you are committing to.

The Power of Physical Exercise

As your device use decreases, something interesting tends to happen: you will have more time. Not all at once, but in small recovered pockets that add up. The average person who reduces their daily screen time by 90 minutes per day recovers over 500 hours of life per year. That is more than 20 days of waking hours.

What you do with that time matters. In the early stages of developing digital sobriety, recovered time can feel uncomfortable, because your nervous system has been accustomed to constant stimulation, and the relative quiet of an unscheduled hour can initially feel like restlessness or boredom rather than freedom. Physical exercise is one of the most effective ways to metabolize that restlessness productively, and it is one of the most reliable tools I know for making sobriety, of any kind, easier to hold onto.

I can tell you from my own life that as my health improved, staying sober became more manageable, not less. The two were never separate projects. When I found a real balance between physical and mental well-being, the cravings lost some of their grip. Watching my own body change, from a softer, more sedentary version of myself into something stronger and more capable, did more for my confidence than any amount of validation I had ever chased on a screen. That kind of confidence is earned in the body, and no notification can counterfeit it.

This is why I want you to start observing your physical state now, here in the Observe step, before we build the actual practice in the Behavior step. You cannot change what you have not honestly looked at.

Here is the part I do not want to gloss over: incorporating physical activity into your S.O.B.E.R. strategy is pivotal, not optional. You do not need to throw yourself into a punishing routine on day one. I began with modest steps and eventually developed a genuine passion for ice skating, of all things. Find what resonates with you. It could be hiking, kickboxing, rowing, cycling, swimming, lifting, or dancing in your kitchen. Embrace something that sparks real enthusiasm and let it carry your recovery forward.

Let me dig into the why, because motivation built on understanding lasts longer than motivation built on guilt. Physical exercise and digital sobriety reinforce each other on multiple levels:

1. **Endorphin release.** Movement prompts your body to release endorphins, your natural painkillers and mood elevators, the source of the so-called runner's high. Those same endorphins ease the anxiety and low mood that so often send your hand reaching for the phone in the first place. You are giving your brain the lift it was trying to extract from the feed, except this version is real and it lasts.

2. **Improved self-esteem.** Regular exercise produces visible, felt changes in your strength, endurance, and physique. As you set and hit small fitness goals, your sense of self-worth grows, and it grows from something you actually did rather than something an algorithm decided to show you. Screen-based validation evaporates the moment you lock the phone. Embodied confidence does not.
3. **Structure and routine.** A standing appointment to move your body gives your day an anchor. For anyone breaking a compulsive habit, predictable routine is grounding, and it crowds out the empty, unstructured moments when mindless scrolling usually creeps in.
4. **Stress reduction.** Physical activity lowers cortisol, your primary stress hormone. By giving stress and tension a physical outlet, exercise reduces the likelihood that you reach for your device to cope. The phone was never actually resolving your stress; it was deferring it. Exercise discharges it.
5. **Improved sleep.** Regular movement contributes to deeper, more consistent sleep, and sleep is the foundation everything else rests on. A rested mind is far better equipped to notice a craving, pause, and choose differently. A depleted mind reaches for the path of least resistance, which is almost always the screen.
6. **Enhanced cognitive function.** Aerobic exercise in particular sharpens attention, memory, and focus, the exact faculties that years of fragmented scrolling have worn down. As you rebuild your ability to concentrate, the shallow pull of the feed has to compete against a mind that is once again capable of deep, satisfying focus.
7. **Connection and support.** Group classes, team sports, a running club, or a standing hike with a friend offer something the phone only pretends to offer: genuine human connection, in person, with people who share a healthy goal. This becomes part of your support network, and it is built on presence rather than pixels.
8. **A replacement, not just a restriction.** Movement is a productive outlet for the restless energy that used to drive you to your phone. Over time, it becomes your new default reach, the thing you turn to when you are bored, anxious, or understimulated. You are not

merely removing a behavior; you are installing a healthier one in its place.

9. **Reconnection with your body.** Compulsive device use pulls you up into your head and out of your physical life. You stop noticing hunger, tension, fatigue, the feeling of weather on your skin. Exercise reestablishes the link between mind and body, restoring a self-awareness and a simple appreciation for being physically here, which is precisely where the phone has been stealing you from.

Exercise is not a cure-all, but few single changes do more to support digital sobriety, and the benefits compound the longer you stay with it.

Now, with the why firmly in place, take an honest inventory. In your SOBER Journal, write down:

- How often do you exercise, and what kind?
- How many hours per day do you spend sitting, and how many of those are spent sitting with a screen?
- How would you rate your current sleep quality?
- When did you last spend a significant block of time outdoors, moving, without your phone?
- What could your body once do comfortably that it cannot do as easily now?
- Is there a physical activity you used to love and quietly let go of?

You do not need to transform your fitness life overnight. Start small, start this week, and start with something you might actually enjoy. In the Behavior step, I will hand you a specific 15-minute, no-equipment routine to launch from. For now, the goal is simply to see your current physical reality clearly on the page, and to begin treating physical presence as the real alternative to digital escape.

Notes

Notes

Behavior

You have done significant work. You have grounded yourself in Stoic philosophy, established your journal practice, completed behavior chain analyses for your specific device triggers, and built an honest picture of your current digital life. Now it is time to take all of that observed, analyzed information and turn it into a new structure for living.

The Behavior step is where insight becomes action. This is the step people often find most immediately gratifying, because for the first time in the process, you are building something new rather than taking apart something old.

Before we move forward, I need you to do something. Step away from your devices, find five minutes of quiet, and ask yourself two questions:

1. Have I been 100% honest with myself in the Stoic and Observe steps?
2. Am I genuinely ready to make specific, measurable changes to my relationship with technology?

If your answer to either question is an uneasy "yes" or a quiet "not quite," I want you to return to Observe. There is no shame in going back. Deeper honesty at the Observe level makes everything that follows more effective. The S.O.B.E.R. Method works only if you work it honestly.

For those ready to proceed: let's build.

Maslow's Hierarchy and Your Digital Life

The Behavior step is about building new structures to live inside, and you cannot build a structure without understanding what it is meant to hold. So before we design a single new routine, I want to introduce you to a map of what a human being actually needs, because every compulsive device habit you have is, underneath, an attempt to meet one of those needs. Understand the need, and you can meet it honestly. Miss the need, and no amount of willpower or app-blocking will hold, because the unmet need will keep reaching for the nearest available relief, which is the phone.

That map is Maslow's Hierarchy of Needs, and I want to give it the time it deserves, because it is the foundation the entire Behavior step is built on.

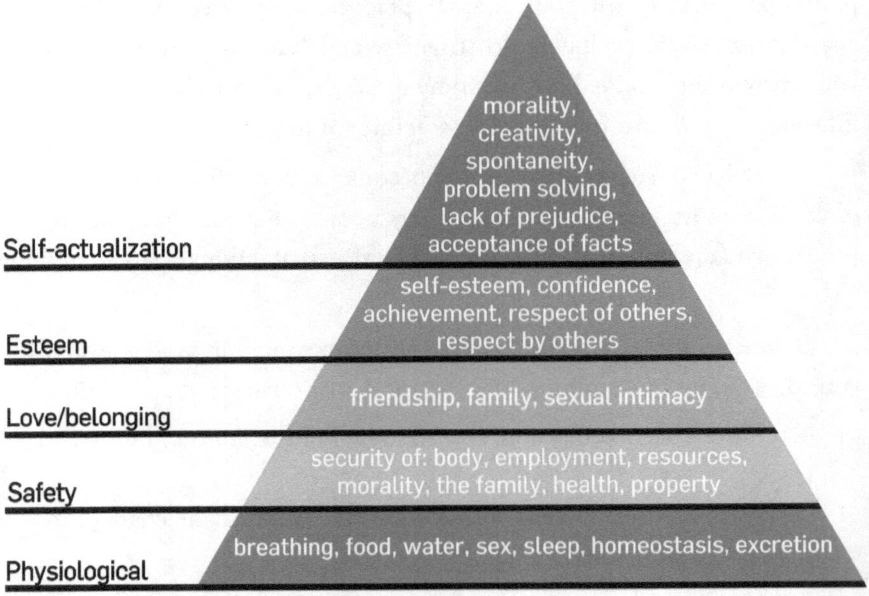

In 1943, a psychologist named Abraham Maslow published a paper titled "A Theory of Human Motivation," in which he proposed that human needs are arranged in a hierarchy, almost always drawn as a pyramid. At the base sit the most fundamental, physical needs. Above them, in ascending order, sit safety, then love and belonging, then esteem, and at the very peak, self-actualization. Maslow's central insight was that the lower needs must be reasonably satisfied before the higher ones can be pursued with any real energy. A person who has not slept and has not eaten is in no condition to worry about self-respect or creative fulfillment. The body's alarms drown everything else out. You build a life the way you build anything sound, from the foundation up.

I have used this hierarchy in every version of the S.O.B.E.R. Method, first for substances and now for devices, because it does two things at once. It shows you, level by level, what a full and flourishing life is actually made of, and it exposes the precise trick your device is playing on you. Here is that trick, stated plainly, because it is the key to this whole chapter: a compulsive device habit masquerades as a shortcut to every level of this pyramid while quietly starving each one. The phone promises belonging and delivers loneliness. It promises esteem and delivers comparison. It promises rest and delivers insomnia. It promises that you are informed and safe while keeping your nervous system in a low hum of alarm. Your entire task in the Behavior step is to stop accepting the counterfeit at each level and start meeting the real need with the real thing.

Let us walk up the pyramid together, one level at a time, and look honestly at what your devices are doing at each.

Level One: Physiological Needs

At the base of the pyramid are the physiological needs, the non-negotiable requirements of a living body: sleep, food, water, movement, and rest. These are not lifestyle preferences. They are the conditions under which a human being can function at all, and until they are met, the brain will not give you reliable access to anything above them.

Your phone does not pretend to feed you or hydrate you. Its damage at this level is subtler and, in a way, more serious, because it attacks the foundation while you are looking elsewhere. The single clearest example is sleep. The late-night scroll delays your sleep onset, the blue light suppresses the melatonin that should be making you drowsy, and the emotionally arousing content keeps your mind running long after the body wants to shut down. Then the cumulative sleep debt impairs the very part of your brain responsible for impulse control, which makes you reach for the phone more the next day, which costs you more sleep, and the foundation crumbles a little further each night. Movement suffers too: hours that could have gone to a body that needs to move instead go to a glowing rectangle that needs you still. Even eating becomes a half-present act performed over a feed.

Meeting this need honestly is the most concrete, highest-leverage work in the book, which is why so much of the Behavior step returns to it: the phone out of the bedroom, the charger in the kitchen, the screen cutoff before sleep, the daily movement. Protect the base of the pyramid first. Everything above it depends on it.

Level Two: Safety Needs

Once the body's immediate needs are handled, the next tier is safety: physical safety, health, a measure of financial security, order and predictability, and, perhaps most importantly for our purposes, freedom from fear. Humans crave a world that feels stable and reasonably under control. Without that baseline sense of security, it is very hard to turn your attention to relationships or growth, because a frightened animal cannot relax enough to flourish.

Here the device runs one of its most convincing cons. The phone presents itself as a safety instrument. It is how you would call for help, how you stay informed, how you keep tabs on the people and events that matter. And some of that is genuinely true. But the same device that could call an ambulance also delivers, around the clock, an algorithmically optimized stream of everything in the world that might threaten you: every disaster, every outrage, every reason to be afraid. Doomscrolling is not an accident, it is the predictable result of feeding a nervous system that evolved to scan for danger an infinite supply of danger to scan. The notifications keep you in a state of low-grade vigilance, never fully at rest. The result is a profound irony: the object you reach for to feel safe and informed is a primary manufacturer of the background fear that makes you feel unsafe.

Meeting the safety need honestly means taking back authority over what enters your mind and when. It means consuming information on a schedule you set rather than from a stream that never ends, silencing the notifications that keep you braced for the next alarm, and allowing your nervous system the long stretches of quiet it needs to actually feel secure.

Level Three: Love and Belonging

Above safety sits our deep, defining need for connection: friendships, family, intimate partnership, a sense of belonging to something larger than ourselves, and the simple experience of being accepted for who we are. We are social animals to the core. When these needs go unmet, loneliness, anxiety, and depression move in. When they are met, we feel a contentment that little else can provide.

This is the level where the device's promise is most seductive and its betrayal most complete. Social media presents itself as connection itself, and in some real ways it does connect us, across distance and time in ways that are genuinely good. But researchers including Robin Dunbar, of "Dunbar's Number" fame, have found that scrolling social media does not provide the same neurological and relational benefits as face-to-face interaction.[17] Passive scrolling in particular, the watching of other lives without truly participating in any, is associated with a *lower* sense of belonging and higher loneliness compared to active, real-world social engagement.[18] You can spend three hours on Instagram "connecting" and come away emptier than when you started. The cruelest part is that the phone does not just fail to feed the belonging need, it actively crowds out the thing that would: it is the screen at the dinner table, the half-attention during a friend's story, the reason a real conversation never quite deepens.

Meeting the belonging need honestly is why the Behavior step puts such weight on phone-free meals, on real conversation, on building an actual support network of people who are present with you. Three hundred followers is not a support network. Three people who will pick up when you call are.

Level Four: Esteem

Near the top of the pyramid are the esteem needs: self-respect, a sense of achievement, recognition from others, a reputation you have earned, and the quiet confidence that comes from mastery of something real. Esteem has an inner face and an outer face. The inner face is your own honest regard for yourself. The outer face is being seen and valued by others. A healthy life needs both, with the inner one load-bearing.

The device offers a brilliant counterfeit of esteem, and it offers it by the minute: likes, hearts, followers, shares, the little numbers that look exactly like recognition. The problem is that esteem built on the reactions of strangers is extraordinarily fragile, because it is entirely outside your control and it resets to zero with every new post. Research from the University of Pittsburgh found that higher social media use was associated with higher rates of depression and anxiety, with the relationship partially mediated by upward social comparison, people feeling worse about themselves after a steady diet of everyone else's curated highlights.[19] You are not building esteem when you scroll, you are renting a feeling of worth and paying for it with your sense of self. The metrics give you a recognition-shaped pellet that never nourishes and always leaves you needing the next one.

Meeting the esteem need honestly means rebuilding it from the inside out, on a foundation no algorithm can touch: the genuine mastery of a skill, the steady achievement of goals you set for yourself, and above all the self-respect that comes from keeping the small promises you make to yourself, day after day. That last one is not incidental. It is the exact mechanism the goal-setting and Execute steps are designed to produce.

Level Five: Self-Actualization

At the summit is self-actualization, Maslow's term for the realization of your own unique potential. This is the level of personal growth and self-discovery, of fulfilling the talents and passions that are distinctly yours, of pursuing a purpose larger than mere comfort, of those rare peak moments of complete absorption and presence, of accepting yourself and others as you are, and of acting from your own internal compass rather than the expectations of the crowd. Maslow was clear that this is not a destination you arrive at and stay, but an ongoing process of becoming.

Self-actualization requires the two things compulsive device use most reliably destroys: sustained attention and autonomy. The deep, absorbed focus that creative and meaningful work demands cannot survive a stream of interruptions, and the capacity for that focus erodes with every year of fragmented scrolling. Just as corrosively, the algorithm quietly relocates your autonomy. When a feed decides what you see, what you want, and what you compare yourself to, you are no longer the author of your own attention, you are a consumer of someone else's design. It is very difficult to become the fullest version of yourself while an optimization engine is choosing your inputs for you. This is the level where the cost of device compulsion is hardest to see and largest in sum, because what it takes from you is not minutes, it is the life you might have made with them.

Meeting this need honestly is, in a sense, the entire point of digital sobriety. When you reclaim your attention and your autonomy, you get back the raw materials of a self-actualized life: the capacity for deep work, the quiet in which purpose can be heard, the presence that lets a peak moment land, and the freedom to choose your own direction. Your Flavor of Confidence lives at this level.

Mapping Your Own Hierarchy

Now make it personal, because this is where the chapter turns from understanding into action. In your SOBER Journal, draw the five levels and work through them honestly, one at a time:

1. What genuine needs are being met in my life right now, at each level?
2. Which needs am I currently trying to meet through my device, and at which level does each device habit really live?
    - Is the late-night scroll a failed attempt at rest, or at belonging?
    - Is the compulsive posting about esteem?
    - Is the doomscrolling about safety and fear?
3. What would happen to that need if I took the device away?
    - And what is the honest, real-world way I could meet it better?

This map is the foundation of your entire Behavior plan. Once you can see which need each device habit is actually, if inadequately, serving, you can design a genuine replacement for it, and the pull of the phone weakens dramatically, because you are no longer asking yourself to give something up. You are giving yourself something better in its place. The needs you identify here become the raw material for the goals we are about to set.

Setting S.M.A.R.T. Date-Certain Goals for Digital Sobriety

This is the engine room of the entire Method, so I am going to slow down and give it the attention it deserves. If the Stoic step gives you the philosophy and the Observe step gives you the truth, this is where the actual change gets manufactured, one small, measurable step at a time. I have watched more people succeed or fail at digital sobriety on the strength of this single skill than on any other. It is the closest thing the Method has to a secret, and the secret is almost embarrassingly simple: small, specific, measured steps, repeated, beat grand intentions every single time.

Here is the principle in one sentence. **The S.O.B.E.R. Method does not run on intentions, it runs on goals, and a goal is just an intention with a date, a metric, and a plan bolted on.**

Vague goals fail. They fail for everyone, every time, and not because you are weak. They fail because a vague goal gives your brain nothing to aim at and no way to know whether it hit. "I'm going to use my phone less" is not a goal. It is a wish. A wish has no edges. You can never quite succeed at it and you can never quite fail at it, which means it can never produce the one thing this entire Method exists to build: the accumulating evidence that you do what you say you are going to do. That evidence is where confidence actually comes from, and you cannot manufacture it with a wish. You manufacture it with a goal you either hit or miss, measured against a date you set in advance.

We turn wishes into goals with a tool called S.M.A.R.T.

What S.M.A.R.T. Means and Why It Works on Devices

The S.M.A.R.T. framework has been used in business and personal development since George Doran laid it out in 1981, and I have leaned on it through every version of this Method, because it does one thing supremely well: it forces a wish to become a plan. S.M.A.R.T. stands for Specific, Measurable, Achievable, Relevant, and Time-bound. Each letter closes a door that a vague goal leaves open for you to wander back out of. Let me walk you through all five in the context of your devices, because the translation is where most people get it wrong.

S, Specific: A specific goal names the exact behavior that will change, in plain language, with no room for interpretation. "Be better with my phone" is not specific. "No phone in the bedroom overnight" is specific. The test is simple: could a stranger read your goal and know exactly what you committed to? "Cut down on social media" fails that test. "Instagram and TikTok only between 5:00 and 5:30 PM, never before, never after" passes it. Specificity is what lets you, and the people holding you accountable, know without argument whether you did the thing.

M, Measurable: This is the heart of the whole Method, so read it twice. A measurable goal attaches a number to the behavior, so that progress stops being a feeling and becomes a fact. Feelings lie. On a bad day you will feel like you have made no progress at all, and the number will quietly tell you that you have cut your pickups from ninety-four a day to forty. On a good day you will feel like you have it beaten, and the number will tell you that you are still scrolling for two hours after midnight. The number is the truth, and your devices hand it to you for free. Your phone is, conveniently, the most heavily instrumented object you own. Screen Time and Digital Wellbeing will give you total minutes, pickups, first-pickup time, notification counts, and per-app breakdowns. Beyond the phone's own data, you can measure nights the phone slept outside the bedroom, consecutive phone-free dinners, days you held your social media window, or minutes from waking to first check. If you cannot measure it, you cannot manage it, and you certainly cannot prove to yourself that you changed it. Pick a metric you can actually read off a screen or count on a calendar.

A, Achievable: An achievable goal is one you can realistically reach in the time you have given yourself. This is where ambition quietly sabotages people. The instinct, on the day you finally decide to change, is to swing for the fences: zero social media, four hours of screen time down to thirty minutes, starting tomorrow, forever. That is not a goal, it is a relapse with a countdown timer. The brain treats an impossible target as a threat and routes around it. Achievable does not mean easy, it means honest. Pushing yourself is good. Setting a deadline you have no real chance of meeting only teaches you, one more time, that you cannot be trusted, which is the exact opposite of what we are here to build. Aim for a step you are about eighty percent sure you can take. Then take the next one.

R, Relevant: A relevant goal is tied to something you actually care about, and to the specific damage you mapped in the Observe step. This is the difference between a goal you borrowed and a goal you own. Do not set a goal because a book told you to, including this one. Set it because your Behavior Chain Analysis showed you that the bedtime scroll is stealing the sleep that makes you snap at your kids, or that the phone at dinner is the reason your partner has stopped telling you about their day. Relevant goals survive hard days, because the reason behind them is real to you. Borrowed goals collapse the first time they are inconvenient. Tie every goal back to a need from your Maslow inventory or a cost from your behavior chains, and write that reason down next to the goal so you can read it when your resolve thins.

T, Time-bound: A time-bound goal has a specific date attached, not "soon," not "eventually," but a day you could circle on a calendar. The date is what converts a hope into a commitment, because a date can be met or missed, and that is precisely what makes it powerful. Date-certain goals are both the guiding stars you steer by and the grounding anchors that hold you steady when cravings and triggers try to pull you off course. "I will leave my phone in the kitchen overnight" is a wish that can drift forever. "By the thirtieth of next month, I will have left my phone in the kitchen for thirty consecutive nights" is a commitment with a finish line.

A goal that satisfies all five is no longer a hope. It is a small, well-defined project, and you already know how to finish a project: you break it into steps and you do the next one.

The Real Secret: Break It Down Until It Is Almost Too Small

Here is the part people skip, and skipping it is the single most common reason digital sobriety attempts fall apart. A S.M.A.R.T. goal still sits at the top of a mountain. The work is in the switchbacks. Large goals feel overwhelming, and an overwhelmed brain reaches for the nearest source of relief, which is, of course, the phone. So you are going to take every goal and break it down into milestones small enough that the next one never feels heavy.

Take a ninety-day goal and work backward. Where do you need to be at sixty days, at thirty days, at the end of week one, and, crucially, what is the one thing you will do tomorrow? A goal of "under two hours of daily screen time in ninety days" starting from a four-hour baseline is not a single leap, it is a staircase: three hours and thirty minutes for the first two weeks, three hours for the next two, and so on, each step a target you are eighty percent sure you can hit. You are not trying to transform your life by Friday. You are trying to take one honest step, prove to yourself you can, and then take the next one. That is the whole game, and it is why the Method works when willpower alone does not.

There is a reason the small steps matter so much beyond just being less scary. Every small step you actually complete deposits a little evidence in an account labeled "I can trust myself." Confidence is not a feeling you summon, it is the interest paid on that account. Miss a giant goal and you withdraw from it. Hit a small one, then another, then another, and the balance compounds until one day you notice you have become a person who follows through. The size of each step is almost beside the point. The streak of kept promises is everything.

Here is the part people skip, and skipping it is the single most common reason digital sobriety attempts fall apart. A S.M.A.R.T. goal still sits at the top of a mountain. The work is in the switchbacks. Large goals feel overwhelming, and an overwhelmed brain reaches for the nearest source of relief, which is, of course, the phone. So you are going to take every goal and break it down into milestones small enough that the next one never feels heavy.

Take a ninety-day goal and work backward. Where do you need to be at sixty days, at thirty days, at the end of week one, and, crucially, what is the one thing you will do tomorrow? A goal of "under two hours of daily screen time in ninety days" starting from a four-hour baseline is not a single leap, it is a staircase: three hours and thirty minutes for the first two weeks, three hours for the next two, and so on, each step a target you are eighty percent sure you can hit. You are not trying to transform your life by Friday. You are trying to take one honest step, prove to yourself you can, and then take the next one. That is the whole game, and it is why the Method works when willpower alone does not.

There is a reason the small steps matter so much beyond just being less scary. Every small step you actually complete deposits a little evidence in an account labeled "I can trust myself." Confidence is not a feeling you summon, it is the interest paid on that account. Miss a giant goal and you withdraw from it. Hit a small one, then another, then another, and the balance compounds until one day you notice you have become a person who follows through. The size of each step is almost beside the point. The streak of kept promises is everything.

A Goal, Built Out, Start to Finish

Let me show you one all the way through, so the abstraction becomes concrete.

The goal, S.M.A.R.T. and date-certain: "By the last day of next month, I will have reduced my daily screen time to under two hours and thirty minutes, measured by the weekly average in Screen Time, with zero phone use after 9:30 PM."

Why it is relevant: My behavior chains showed the late-night scroll is costing me an hour of sleep, and I am exhausted and short-tempered by Wednesday.

The milestones: Week one, charger moves to the kitchen and a book goes on the nightstand, no screen-time target yet, I am only changing the environment. Week two, daily average under three hours and thirty minutes, hard stop at 9:30 PM at least five nights. Week three, under three hours, 9:30 PM stop six nights. Week four, under two hours and thirty minutes, 9:30 PM stop every night.

Tomorrow's single step: Tonight, move the charger and put the book on the nightstand. That is it. That is the whole task for day one.

Notice that the first milestone barely asks anything of your willpower. It changes the setup, not the struggle, which is exactly how you want to begin. Notice too that every step is measurable, so at any point you can open Screen Time and know, as a fact and not a feeling, whether you are on the staircase or off it.

Measuring Your Progress

Setting the goal is half of it. Measuring is the other half, and it is not optional, because a goal you do not track is just a wish wearing a date. Build these into your weekly rhythm.

Track regularly. Pick a set time each week, I suggest a few minutes every Sunday evening, to write your numbers in your SOBER Journal. Weekly screen-time average, daily pickups, nights the phone slept outside the bedroom, whatever your metric is. Same time, every week, no exceptions.

Use your tools. Screen Time and Digital Wellbeing do the counting for you. Your job is only to read the number and record it by hand in the journal, where you will actually confront it. I remain ambivalent about fighting a device problem with the device, so keep the recording analog even though the data is digital.

Celebrate small wins. Every milestone you hit, no matter how minor it looks from the outside, is real. Thirty phone-free dinners is an achievement. Mark it. Acknowledging the small wins is not self-indulgence, it is what keeps the account of self-trust visibly growing, and that visibility is fuel.

Adjust as needed, and remember the rule. If you keep missing a milestone, stop and ask why. Maybe the step was too big, maybe a trigger you did not anticipate is in the way, maybe the plan itself needs a rework. Change the plan freely. But here is the rule I want you to tattoo somewhere in your mind: change the plan, never the goal. The plan is a route. The goal is the destination. When the road is washed out, you find another road, you do not pick a closer destination because the first one got hard.

Seek feedback. Where you can, let an accountability partner see your numbers. An outside perspective catches the rationalizations you cannot see in yourself, and saying your progress out loud to another person makes it real in a way a private journal cannot.

Visualize your progress. Make it physical and visible. A simple wall chart, a calendar where you mark each phone-free night, a checklist on the refrigerator. Seeing the chain of kept promises lengthen in front of you is one of the most motivating forces available to you, and it costs nothing.

Reflect and analyze. At regular intervals, step back and read the pattern, not just the day. What is working? What keeps tripping you? Which triggers are losing their power and which are digging in? Those patterns are the raw material for your next behavior chain and your next goal.

Why Date-Certain Goals Carry So Much Weight

It is worth naming plainly what these goals are doing for you, because understanding the mechanism makes you more likely to honor it. A date-certain goal gives you tangible milestones to aim at in what can otherwise feel like a formless effort. It supplies motivation, because an approaching date sharpens resolve. It creates accountability, because a fixed date is a promise you made to yourself in writing. It makes your progress measurable, so you can see how far you have come and not just how far is left. It even works as a coping mechanism: when a craving hits, returning your attention to the specific goal and its date is a place to put your mind that is not the phone. It gives you defined points to celebrate, which keeps morale up. And it guarantees forward momentum, because the moment you achieve one goal you set the next, and the cycle carries you onward. Guiding stars and grounding anchors, both at once.

How to Set Yours

Now build your own. Work through this in your SOBER Journal.

1. **Mine your behavior chains.** Your BCAs already identified your specific triggers and compulsive behaviors. Choose the three that do the most damage to your daily life, your relationships, or your self-respect. Start there, not with whatever feels easiest.
2. **Write the target behavior, specifically.** For each of the three, name the exact behavior you want in its place. Not "less Instagram," but "Instagram only during my 5:00 to 5:30 PM window."
3. **Run it through S.M.A.R.T.** Make each goal Specific, Measurable, Achievable, Relevant, and Time-bound. If any letter is missing, the goal is not finished. Write the relevance, the reason it matters to you, right next to it.
4. **Set the date, then build the staircase.** Thirty days for an initial change, sixty to ninety for a habit to consolidate, twelve months for a full transformation. Then break the span into weekly milestones, and name tomorrow's single small step.
5. **Choose your metric.** Screen-time average, pickups per day, nights without the phone in the bedroom, consecutive phone-free dinners, days under your social-media limit. Pick something you can read or count without guessing.
6. **Give each goal its own page.** Goal at the top, the reason it matters beneath it, the date, the metric, and the milestone dates down the side with room to record your weekly numbers against them.

Do this, and you no longer have a wish, or even three wishes. You have a project plan, with a finish line, a measuring stick, and a next step you can take tonight. And a project plan, unlike a hope, is something a human being actually knows how to complete.

Building a New Daily Routine

I have always believed in routines. Not because life is predictable, it is not, but because structure is what allows genuine freedom. When the foundational decisions of your day are made in advance and run on positive autopilot, your conscious attention is available for things that actually require it.

Here is the reality of digital sobriety routines: the most powerful changes often happen in the first hour after waking and the last hour before sleep. These are the two periods when device habits are most entrenched and when changing them yields the most dramatic benefits.

The practice of no phone first thing in the morning, which means getting out of bed, making coffee, eating breakfast, exercising, and beginning your day without looking at your phone for the first 60 to 90 minutes, is one of the highest-leverage changes documented in digital wellness research. A 2025 randomized controlled trial published in *BMC Medicine* found that reducing daily smartphone screen time produced measurable improvements in mood, stress, sleep quality, and overall well-being, with the benefits fading once participants drifted back to their old levels of use.[20] The first thing you consume cognitively in the morning sets the emotional weather for the hours that follow. Choosing what that is, rather than letting the algorithm choose, is a foundational act of self-determination.

Similarly, the last hour before bed should be phone-free. Blue light from the screen suppresses melatonin, the hormone that helps you fall asleep. At the same time, social content keeps the mind alert and emotionally engaged, making it harder to wind down. Your brain also uses sleep to process and consolidate memories. The quality of that work depends on what you gave it to work with in the hours before you lie down.

Here is a sample device-aware daily routine. Adapt it to your actual life:

» **5:30 - 6:00AM:** Wake up. Make bed. **No phone.**

» **6:00 - 7:00AM:** Physical movement (walk, workout, yoga, skating, or whatever you love). **No phone.** If you use music, download it before bed so you are not unlocking.

» **7:00 - 8:00AM:** Shower, breakfast, ten minutes in your SOBER Journal. Read one passage from Meditations. **Still no phone.**

» **8:00 - 8:30AM: First intentional phone check** of the day. Respond to anything urgent. This is not passive scrolling; this is a purposeful, time-bounded check.

» **8:30AM - 12:00PM:** Deep work or primary obligations. **Phone on Do Not Disturb or in another room.** Notifications off except for calls.

» **12:00 - 1:00PM:** Lunch. If with others, phone away from table.

» **1:00 - 5:00PM:** Continue work. Scheduled communication checks at 2:30PM and 4:30PM.

» **5:00 - 5:30PM:** Intentional social media window if applicable. Scheduled and time-bounded, not passive scrolling but intentional engagement.

» **5:30 - 6:00PM:** Review your S.O.B.E.R. Method journal entries. Note progress.

» **6:00 - 9:00PM:** Personal time, meals with others, physical activity, connection. Phone present but not primary.

» **9:00 - 9:30PM:** Final check of messages if needed. Write in SOBER Journal. Review your BCAs.

» **9:30PM: Phone to kitchen.** Charged there overnight. Book on nightstand.

» **10:00PM:** Sleep.

This is an example, not a mandate. The essential structural elements are: no phone first thing, no phone before bed, scheduled rather than reactive communication checks throughout the day, phone-free meals, and physical movement daily. Build from there.

Notice that I included review of your Behavior Chains in the evening. This is purposeful. Your BCAs are not static documents; they are living analyses that you update as your behavior evolves. New triggers will emerge. New vulnerable moments will appear. Old triggers will lose their power. The nightly review keeps you conscious of the chains you are still working to modify and gives you data for your journal.

The Notification Architecture Redesign

Here is an uncomfortable fact: your notifications are not organized for your benefit. They are organized for the benefit of the apps that send them.

Every notification that interrupts you is a small act of enrollment in someone else's revenue model.

The average smartphone user receives between 65 and 80 notifications per day. Research from multiple sources, including a comprehensive study by Kushlev and Dunn published in *Computers in Human Behavior,* found that reducing notification frequency improved focus, reduced feelings of stress, and improved overall mood and perceived productivity.[21] Simply reducing how many times per day your phone demands your attention changes your cognitive experience of the day.

Here is your notification redesign protocol:

Step 1: Audit every notification. Go to Settings → Notifications on your device. Look at every app that has notification permissions. Ask yourself: Does this app need to reach me in real time? What would actually happen if I saw this notification 6 hours later instead of now?

Step 2: Classify by neccessity. Keep notifications active for:

» Phone calls (real people, real urgency)
» Direct text messages from people you genuinely care about
» Calendar reminders
» Safety and emergency alerts

Turn notifications off entirely for:

» Every social media app
» Every news app
» Every game
» Every shopping or retail app
» Email (check email on your schedule, not email's)
» Most productivity apps

Step 3: Disable badges and sounds for most apps. Even if you leave an app with some notification access, removing the red badge numbers and sounds substantially reduces the Pavlovian pull of those notifications.

Step 4: Consider one week of total notification silence. For an experimental period, turn off all app notifications and see what you miss. Almost everyone finds they missed nothing of consequence, and that the main thing they gained was the ability to decide when to engage rather than being interrupted.

Grayscale Mode

Grayscale mode is one of the most counterintuitive and well-supported behavioral interventions for device overuse. Converting your phone display to grayscale, which means showing everything in shades of gray rather than color, sounds minor. The effect is not minor.

Human visual attention is strongly drawn to color. The reds and oranges of notification badges, the vibrant imagery of social media feeds, the warm yellows and greens of certain apps: all of these are designed with knowledge of how color captures and directs attention. Remove the color, and the visual pull diminishes substantially.

A 2024 randomized controlled trial published in *PLOS ONE* by Andrews and colleagues found that participants who switched their phones to grayscale mode reduced their daily screen time by an average of 24 minutes per day, without any other behavioral intervention.[22] Participants also reported feeling greater sense of control over their device use and lower perceived stress from notifications. Twenty-four minutes per day is not trivial: over a year, it represents approximately 146 hours of recovered attention.

The grayscale effect works through several mechanisms:

» Apps designed primarily for visual consumption, like Instagram, Snapchat, TikTok, and YouTube, become significantly less viscerally appealing in gray.

» The phone as a functional tool, only for calls, texts, and navigation, loses almost no utility in grayscale.

» Each time you unlock the phone and see the gray screen, it creates a brief moment of conscious awareness: "I've made an intentional choice about my device use," interrupting the autopilot just long enough for the prefrontal cortex to participate.

How to enable grayscale:

» **iPhone: Settings** → Accessibility → Display & Text Size → Color Filters → Enable with Grayscale

» **Android: Settings** → Digital Wellbeing → Bedtime Mode → Grayscale or Monochromacy (or search Settings for "Grayscale")

I recommend starting with grayscale during your highest-vulnerability periods: the first thing in the morning, the last hour before bed, during meals. Some people extend it to their entire day.

Device-Free Zones and Times

Physical separation from devices is one of the most consistently supported interventions in digital wellness research. The research by Adrian Ward and colleagues at the University of Texas at Austin found that having your smartphone on your desk, even face-down and even silenced, measurably reduces available cognitive capacity, because part of your brain is continuously devoting resources to not checking it.[23] The phone does not have to be in your hand to occupy your attention. It just has to be nearby.

This finding has a clear practical implication: when you want your full cognitive presence somewhere, the phone cannot be in the room.

Establish the following device-free zones in your life:

The dining room or dining table: This one I feel strongly about and personally enforce. The average meal takes fifteen minutes. Fifteen minutes of genuine, phone-free conversation with someone you care about. These moments are irreplaceable, and they slip away faster than any of us expect. I still remember specific dinnertime conversations from my childhood. I do not remember anything I saw on a screen during those same years.

The bedroom: The research on sleep quality and bedroom devices is unambiguous. If your phone is not in the bedroom, you cannot reach for it in the night, and you will not wake up with it already in your hand. Get an old-fashioned alarm clock if you use your phone as an alarm. The modest cost of that clock is one of the highest-return investments in your life quality you can make.

The first 30 minutes of your workday: Give your brain the opportunity to arrive at work before your inbox does.

Conversations: When you are in a meaningful conversation, whether a one-on-one with someone you care about, a performance conversation with a direct report, or a therapy session, your phone should not be visible. Not just silenced. Not just face-down. Not in the same room.

Building Replacement Behaviors

The most common reason device reduction efforts fail is not lack of motivation. It is the vacuum. You remove the phone, and there is nothing to do with your hands, your eyes, and your nervous system's need for stimulation.

You need replacement behaviors. Not forced, dutiful activities; these should be behaviors you actually want to engage in, that meet the same underlying need the phone was meeting, but more completely.

Here is a framework:

For boredom: The phone meets boredom through passive stimulation. Better replacements are physically present activities: a physical book, a walk, a musical instrument, a craft, a conversation with a real human, cooking something from scratch. The key characteristic is that they require presence rather than allowing the dissociative partial-attention that scrolling produces.

For anxiety: The phone meets anxiety through distraction. Better replacements are practices that process the anxiety rather than defer it: physical exercise, journaling, breathing exercises, meditation, calling someone directly rather than passively monitoring their social feed. The 5-4-3-2-1 grounding technique, which asks you to identify five things you can see, four you can touch, three you can hear, two you can smell, and one you can taste, is immediately accessible and surprisingly effective for acute device-craving anxiety.

For social connection: The phone meets the need for connection through social media and messaging, but research consistently shows these forms of connection leave people feeling less satisfied than alternatives.[17] Better replacements: scheduling regular in-person meetings, phone calls (voice calls), and handwritten correspondence. I have a structured monthly gathering with a core group of close friends, supplemented by a group text that is sometimes busy and sometimes silent for weeks. That is a real support network. Three hundred Instagram followers is not.

For entertainment: Watching content on your phone during solitary time is legitimate, but the difference between intentional entertainment, such as choosing a film, watching it, stopping, and passive scroll-entertainment, such as opening TikTok and coming up forty-five minutes later, is the difference between a deliberate pleasure and an escape that leaves you feeling vaguely worse. Choose your entertainment. Don't let it be chosen for you by an algorithm.

Exercise Plan

In the Observe step, you inventoried your current physical state. Now you are going to begin building the physical practice that will support your digital sobriety. Exercise is not a side note here; it is a load-bearing element of the Method.

You do not need a gym membership. I think gym memberships are one of the most reliably wasted expenses available to the modern consumer. What you need is a minimum of 30 minutes of vigorous physical movement five times a week, and a growing set of activities you actually enjoy.

The reason physical exercise is so important to digital sobriety specifically: your phone is, among other things, a sedentary comfort object. The more physically active your life is, the less attractive sedentary phone scrolling becomes. Additionally, exercise produces the neurochemical rewards of dopamine, endorphins, and serotonin that your brain has been getting from device notifications. You are literally building an alternative neurochemical supply.

Here is a 15-minute phone-free body-weight routine you can do in your living room, requiring no equipment. Do each element at full intensity for 45 seconds, rest 15 seconds, then move to the next:

» **Warm-Up**

- Jumping Jacks: Stand with feet together and arms at your sides. Jump your feet apart and raise your arms overhead. Reverse the movement and repeat.

• Lunge with an Explosive Knees: Step forward with one leg and lower your body until both knees are bent at a 90-degree angle. Powerfully push through the heel of your front foot and use that momentum to drive the back knee forward and up. The movement should be explosive, as if you are trying to bring that knee towards your chest. For added intensity, you can add a small jump with the front foot as you drive the knee forward. Switch forward leg, repeat while continue to rotate legs.

» **Cardio Boost**

• Burpees: Begin in a standing position, then squat down, place your hands on the ground, and jump your feet back into a plank position. Jump the feet back towards the hands, and explosively jump into the air. Repeat.

• Mountain Climbers: Begin in a plank position and rapidly bring one knee towards your chest at a time, as if you're climbing.

» **Strength and Conditioning**

• Push-Ups: Start in a plank position with arms straight. Lower your body until your chest almost touches the floor. Push through your arms, returning to the starting position. Start on your knees, if necessary, until you build enough strength.

• Squats Punch: Stand with feet shoulder-width apart. Lower into a squat position, keeping your knees in line with your toes. Push through the heels to return to a standing position and as you do cross-body punch. Alternate arms every time you stand.

• Plank Twist: Hold a plank position, maintaining a straight line from head to heels. Engage your core, rotate your hips to the left, return to center, rotate to the right, return to center, repeat.

• Squat Jacks: This is a combination of a normal squat and a jumping jack. Start by doing a normal squat and as you rise do a jumping jack, then return to a squat. Repeat.

» **Core Work**

- Sit-Ups with Punch: Lie on your back with knees bent. Bring your fists together at your mid-chest. Engage your core and lift your upper body off the ground. As you sit up, cross-body punch, lay back down, sit up again, cross-body punch the other direction, repeat while alternating punch arm.
- Leg Raises with Alternating Legs: Lie on your back with legs straight. Lift both legs slightly off the floor. Now lift your left leg off the ground to about a 45-degree angle, then lower without letting it touch the floor, lift the right leg, lower, then lift both legs. Repeat.
- Russian Twists: Sit on the floor, lean back slightly, and lift your legs off the ground. Hold your hands together and twist your torso to the left, then to the right, tapping the floor beside you.

» **Cool Down**

- Forward Bend: Stand tall, exhale, and hinge at the waist, reaching for the ground. Hold for 20 seconds, roll up, repeat.
- Quad Stretch: Stand tall and pull one foot towards your glutes, holding your ankle. Keep knees together and push your hips forward. Hold for 15 seconds on each leg, 2 reps.
- Child's Pose: Kneel on the floor, sit back on your heels, and extend your arms forward on the ground. Rest your forehead on the ground and hold for 1 minute.

Remember, it's essential to keep your form correct to prevent injuries and maximize the benefits. Stay hydrated, and listen to your body. If a certain exercise causes pain, beyond typical muscle fatigue, adjust or skip that move. As you progress, increase the speed and intensity or modify steps to continue challenging yourself. For example, there are over 50 types of Mountain Climber exercises. Change it up to work different areas of your body. This is just a great example. Feel free to build your own. Try to start with every other day. If you can make it part of your daily routine, that would be fantastic.

Dietary Plan for Digital Detox Support

Nutrition is rarely the first thing people think of when they set out to change their relationship with a screen, and I understand why it sounds like a detour. Stay with me, because it is not. The body you are trying to bring back into your own life is the same body you feed three times a day, and what you put into it directly shapes the moods, the cravings, and the impulse control you are going to be drawing on every hour of this work. In every version of the S.O.B.E.R. Method I have ever taught, diet has been a load-bearing pillar, not a footnote, and it is no different here. Take it as seriously as you take the journal.

Let me start with the biochemistry, because it explains why this matters so much for device addiction specifically. Problematic device use dysregulates dopamine. The rapid, variable hits from notifications, likes, and the next swipe train your brain on a reward schedule it was never built for, and over time the brain protects itself by turning down its own dopamine sensitivity. The technical name is dopaminergic downregulation. The practical result is that ordinary life, a conversation, a walk, a meal, a quiet afternoon, begins to feel flat and gray by comparison, and the brain increasingly needs the device just to feel normal. This is the same fundamental machinery that drives substance addiction, and it means that during your early weeks of digital sobriety your reward system is, in a real sense, recalibrating. What you eat either supports that recovery or fights it.

Mindless Scrolling and Mindless Eating Are the Same Disease

Before the food itself, I want to name something I have seen in nearly everyone I have worked with, myself included. The compulsion to scroll without noticing and the habit of eating without tasting are not two separate problems. They are the same problem wearing two outfits: the disease of absence, of being somewhere other than where your body actually is. Think about how most snacking happens now. A bag of something in one hand, the phone in the other, eyes on the feed, and twenty minutes later the food is gone and you could not describe a single bite. You were not eating. You were feeding a hand while your attention was somewhere else entirely.

This is why one of the simplest pillars of this Method, the phone-free meal, does double duty. When you take the phone off the table, you do not just recover the conversation and the presence. You start tasting your food again. You notice when you are full, because you are actually in the room with your own body when the signal arrives. Mindful eating and digital sobriety turn out to be the same skill practiced on different objects: the skill of being present enough to make a real choice instead of running on autopilot. Heal one and you tend to heal the other.

There is a portion lesson buried in that, and it is the same one I have given for years. You do not need to count every calorie. You need to be present enough to stop when your body tells you to. A rough guide I have used and shared: a portion of any one food about the size of your fist, which is roughly the size of your stomach, is usually plenty. The first weeks of eating that way feel like deprivation, and then your stomach quietly adjusts and the constant low hum of wanting more fades, in food exactly as it does with the phone. Presence is the through-line. It always is.

The Plan

You did a first inventory of your physical state in the Observe step. Now build the eating habits that will support everything else you are constructing. None of this is exotic, and that is the point. It is simple, sustainable, and aimed squarely at steadying the brain chemistry your device habit has knocked off balance.

1. **Stay hydrated.** You have heard it a thousand times and it still bears repeating: drink water throughout the day. Even mild dehydration amplifies anxiety and degrades decision-making, which are precisely the conditions under which the hand reaches for the phone without asking.
2. **Eat regular, phone-free meals.** Three balanced meals at reasonably steady times stabilize your blood sugar, and stable blood sugar means fewer mood swings and fewer cravings, including the craving to

escape into a screen. Erratic eating and erratic scrolling feed each other. Steady the one and you help steady the other. And every one of those meals is eaten with the phone in another room.

3. **Build meals on tyrosine-rich proteins.** Chicken, turkey, fish, eggs, and legumes supply the raw material your body uses to synthesize dopamine. As your reward system recalibrates, give it what it needs to make its own dopamine instead of begging the phone for a hit.
4. **Choose complex carbohydrates over refined ones.** Swap white bread and sugar for oats, quinoa, brown rice, and sweet potatoes. They support serotonin production and deliver steady energy instead of the spike-and-crash that leaves you depleted and reaching for stimulation.
5. **Fill half the plate with fruits and vegetables.** The vitamins, minerals, and antioxidants here support mood, immune function, and overall brain health. This is unglamorous advice that quietly does more than almost anything else on the list.
6. **Include healthy fats.** Avocado, nuts, seeds, and oily fish provide the omega-3 fatty acids that reduce neuroinflammation and support mood regulation. A brain rebuilding its reward circuitry runs better on these.
7. **Get enough magnesium.** Dark leafy greens, seeds, and a little dark chocolate support stress regulation and sleep, the same sleep your device habit has been stealing.
8. **Feed your gut.** Fermented foods such as yogurt, kefir, kimchi, and sauerkraut support the gut-brain axis, which research increasingly links to mood and anxiety. Your gut is part of your nervous system. Treat it like it matters.
9. **Cut the sugar and ultra-processed foods**, and notice the parallel. Refined sugar produces its own dopamine spike followed by a crash that leaves you worse off and wanting more, which is the notification loop in edible form. The same brain pattern you are working to interrupt on your phone is being reinforced three times a day at the table if you let it. Interrupt it in both places.

10. **Mind your caffeine.** When you pull stimulation out of your day, the temptation is to pour another stimulant in to fill the gap. Be careful here. Too much caffeine amplifies anxiety and wrecks the sleep that your impulse control depends on. You are trying to calm an overstimulated system, not trade one jittery loop for another.
11. **Listen to your body.** As you make these changes, pay honest attention to how different foods leave you feeling. Some lift you, some flatten you, and your body, when you are present enough to hear it, will tell you what it needs. I have learned my own signals over the years. I get a powerful craving for dark chocolate before a depressive mood swing, and I can often blunt the swing by heeding it early. The point is not my particular tell. The point is that the same attentiveness digital sobriety is teaching you, the willingness to notice what is actually happening inside you, is the same attentiveness that makes you well fed.
12. **Seek professional advice where you can.** You are one of a kind, and so are your nutritional needs. If it is within your means or covered by your insurance, a nutritionist or dietitian can tailor this far better than any book. And as the disclaimer at the front of this book says plainly, I am not a medical professional. Treat anything here that touches your health as a starting point for a conversation with someone qualified, not as a prescription.

Recovery is holistic. You cannot scroll your way out of device addiction and you cannot will your way out of it either, not while running your body on sugar, caffeine, and skipped meals. Nourishing yourself properly is not separate from the work. It is the ground the work stands on, and the steadier that ground, the lighter every other part of this Method becomes.

Your Network of Sober Friends

What I have personally found most powerful in my own life is not an app or a program; it is people. Specifically, a structured network of people who are present with me, who challenge me, and who are honestly invested in each other's growth.

Build your network for digital sobriety deliberately:

1. **Tell someone you trust what you are doing.** Have them read this book. Ask them to hold you accountable. Having one person who asks you, genuinely, "How is the phone thing going?" changes the internal experience of the commitment.
2. **Find or create a regular social gathering that is explicitly phone-free.** It can be a monthly dinner, a weekly walk, a standing coffee date. The explicit phone-free nature is not a rule so much as a shared commitment: we are here for each other.
3. **Identify the people in your life who model healthy technology use.** They exist. Spend more time with them. Observe what they do differently.
4. **Join the free online community at sobermethod.org.** It is just getting started, which means you will not be joining a crowd; you will be helping to build something from the ground up. Introduce yourself. Find others working the Method for digital sobriety. And then, the moment you are able, use it for its real purpose: to find people near you and start meeting in person. An online room is a fine place to find your people. It is a poor place to keep them. Take it offline. Walk together, eat together, sit in a room together with the phones in a basket by the door. That is where the actual rebuilding happens.

Now, let me name the obvious thing before you do, because you are too smart not to notice it and I would rather not look like a hypocrite. Yes, I am pointing you to an online community as part of a program to heal your relationship with online life. I see the irony. I am leaning into it on purpose. The goal of this book was never to make you flee your devices in shame. It was to put you back in charge of them, to make your technology a tool you reach for deliberately rather than a reflex that reaches for you. Using an online space, consciously, with a clear purpose, and as a bridge to real-world connection, is not a betrayal of that goal. It is the goal, practiced. I am simply meeting you where you already are, so that together we can get you somewhere better, on your own terms.

Notes

Notes

Notes

Execute

You have done a great deal of work on yourself to arrive at step E of your S.O.B.E.R. journey. You have grounded yourself in Stoic philosophy and named the virtues you most need to develop. You have observed your actual device behavior with more honesty than most people ever bring to this challenge. You have designed concrete changes, built a new daily routine, and set specific, measurable, date-certain goals. The primary objective of this step is simple to say and harder to do: you are going to implement all of that planning. You are quite literally going to execute your daily life in alignment with the goals you carefully created.

This step can feel a little daunting if you do not think of yourself as the most organized person in the world. Do not start by setting a limiting belief about yourself. I have watched some of the most disorganized people I know follow their S.O.B.E.R. plan and astonish themselves. You can do this.

Before we go any further, I need you to pause. Clear your mind, step away from the distractions, find some silence, and give yourself five uninterrupted minutes. During that time, ask yourself two questions:

1. Have I been 100% honest with myself in the Stoic, Observe, and Behavior steps? The S.O.B.E.R. Method requires complete honesty with yourself. You cannot journal one thing and quietly believe another. Without that honesty, you will struggle, and you will likely fail quickly.
2. Am I ready to commit to the difficulties ahead, to change my life for the better, and to find my Flavor of Confidence in the digital world?

If you answered no to either question, or if you were an uneasy yes, I want you to return to the beginning of Observe. Re-evaluate your thinking and dig deeper within yourself. There is no shame in it. Deeper honesty now is exactly what makes everything that follows actually work.

Before you decide, I want to commend you for the work you have already done. You are amazing. Digital sobriety, like any sobriety, is a deeply personal journey, and until you are truly ready, you will not fully embark on it. Here is the truth: despite all of your work so far, despite the interventions and the support of the people who love you, until you firmly decide for yourself that it is time for a change, that change will remain just out of reach. The external pressures, however well-meaning, are not enough. You have to look inward and decide, for yourself, that it is time to address this.

You will feel it inside, with the firmest resolution. For me, that resolution did not come easily. After my near-death experience, I relapsed in three weeks. Then I went to rehab, after which I relapsed in multi-day binges and detoxed in a yo-yo fashion seven times over the next three months. I put my body through an incredible amount of pain in those detoxes, and I faced some frightening hallucinations as I withdrew, in very public fashion, on the street. Then I woke up on the morning of March 23, 2016 with a life-altering moment of clarity: you have to stop right now, or you are going to be dead in a month. I picked up the phone, called Samantha, and told her I was going to live the rest of my life sober. I have. Not that I ever feel an urge today, yet knowing that Samantha is looking down on me at all times definitely provides an unusual check on my actions. I mean, what if she has the power to smite me? You may chuckle. I say that in good humor.

What drives you from within is almost always more potent than anything outside you. Family and friends may want the best for you, but the desire for change has to be a flame lit from the inside. When the drive is internal, the commitment is stronger and the resolve to see it through is firmer. It is easy to let others down. It is much harder to let yourself down.

I am pulling for you. Your sobriety means a great deal to me. Those are not hollow words from an author. They come from the heart of someone who stood face-to-face with the dragons of his addiction, who lost the person who meant the most to him at the hands of a drunk driver, who has committed the rest of his life to helping others find their sobriety, and who is driven to build a global network of S.O.B.E.R. Method Chapters so that anyone, anywhere, can find the local support they need. Your success is my success. Find that strength deep within you. It is there. I see it.

For those ready, let's proceed.

As I have walked my own path, I have come to appreciate the wisdom of Dr. Jordan B. Peterson, the renowned psychologist and author. He often emphasizes that "not setting goals is like living your life in a fog." He could not be more correct. Without clear, well-defined goals, it is nearly impossible to hold yourself accountable, and accountability is the entire engine of this step. In the Behavior step, you reflected on your behavior chains and set date-certain goals. Now I want to give you some practical guidance on how to build a daily plan to execute those goals and measure your progress. These are not skills they teach you in primary school. If you are not a project manager, it is unlikely you have ever had reason to build a calendar plan to accomplish a goal. To all the project managers reading this, I apologize in advance for the crudeness of my explanation. Consider this an entry-level class, and if you find a better way to teach it, I would genuinely welcome your ideas. Please reach out.

Building Your Project Plan

Open up your SOBER Journal and select one of your goals. Flip to a blank page, just for that goal. Re-write the goal at the top of the page, and in the top left corner, write the date you have set for it. You will repeat this process for each of your goals, again and again, as you cycle through the S.O.B.E.R. Method in the years to come. What you are building here is a project plan, and a project plan is what separates the people who change from the people who only ever mean to.

1. **Identify your end goal.** Review what you want to achieve, and be specific. Not "I want to be on my phone less," but "I want to average under two hours of daily screen time for thirty consecutive days." If you did the work in the Behavior step, your goal is already specific.
2. **Set a clear date.** This is the date-certain part. Decide the exact date by which you will achieve the goal, and mark it on a calendar in bold. "I will reach thirty straight days under two hours by the last day of next month."
3. **Break it down.** Big goals feel overwhelming, and an overwhelmed

mind reaches for the phone. Divide the end goal into smaller, manageable milestones: perhaps "phone out of the bedroom for thirty nights," then "daily average under three hours," then "under two hours and thirty minutes," and on toward the target.

4. **Assign dates to milestones.** Give each milestone its own date, just as you did the end goal. Shorter targets with their own deadlines keep you moving and keep the finish line from feeling impossibly far away.

5. **Define measurable metrics.** Decide exactly how you will measure progress: daily screen-time average, number of pickups, nights the phone slept outside the bedroom, consecutive phone-free dinners, days you held your social-media window. Choose a metric for every goal so you can see your progress at a glance.

6. **Document everything.** Build a visual project plan. Your SOBER Journal is my recommendation, though a simple spreadsheet works, and the project managers among you will drop it into your favorite tool. Put the end date, the milestones with their dates, and the metrics all in one place. This is your roadmap.

7. **Regularly review and adjust.** Set aside time each week to review the plan. Check off the milestones you have hit. Watch your metrics. If you are off track, adjust the plan, not the goal. Maybe you need more support, a smaller next step, or a revised milestone. Lean on your accountability partner or mentor. Keep your eyes on the goal.

8. **Celebrate milestones.** Do not wait until the finish line to acknowledge yourself. Every milestone you reach is proof you are making progress and fuel to keep going. Mark it.

9. **Seek feedback.** Share your project plan with someone you trust: your mentor, your accountability partner, or a friend doing this alongside you. An outside eye catches what you cannot, and offers encouragement when you most need it.

10. **Stay flexible.** Life is unpredictable. Milestones may need adjusting, and new challenges will arise. That is fine. The plan exists to guide and motivate you, not to box you in. Adjust as needed,

always keeping the end goal in sight.

Remember, the journey toward any significant achievement is often just as important as the achievement itself. By turning your date-certain goals into a written project plan, you are handing yourself a roadmap to success. Stay committed, stay flexible, and trust in your ability to reach the goal.

The Daily Journal as Your Execution Tool

Having a project plan for each goal is a great start. From here, the daily recording of your progress is what actually carries you. It gives you the data to steer by, and it gives you something else too: a few honest data points to share with the people walking beside you. Everyone wants to get behind a good story, and the one you are writing, the story of a person taking back their attention and their life, is a good story.

In the Observe and Behavior steps, your SOBER Journal was mostly a place for reflection. In Execute it becomes your primary measurement instrument as well. Each day, begin with the hard numbers: your total screen time, your top apps, your pickup count, and a one-line status on each goal, on track or off track. Record them by hand, even though the phone already counted them, because writing a number down is what makes you actually confront it. Then go deeper. Here is how to make the most of your journaling.

1. **Use your SOBER Journal.** I had you choose your favorite notebook, and I stressed handwriting for a reason. I hope you took that seriously, because a journal you genuinely want to open is a journal you will actually use every day. Keep it analog, and keep it away from the device.
2. **Set aside consistent time.** Make journaling a daily ritual, whether first thing in the morning to set your intentions or last thing at night to reflect. It should already be sitting in the daily routine you built in the Behavior step. A few minutes, the same time, every day.
3. **Document your feelings.** Note how you feel. Strong today? Restless? Fighting the pull to check? Write the highs, the lows, and everything between. Over time this reveals your patterns and your triggers, and you will use those notes to modify your behavior chains.

When a new trigger situation appears, star it in the journal as an additional BCA to run on your next cycle. This is the continuous-improvement heart of the S.O.B.E.R. Method.

4. **Log any temptations.** Did something pull you toward the phone today? Write it down. What was the trigger? How did it feel? And, most importantly, how did you respond or cope? This is how you devise better strategies for next time.

5. **Track your achievements.** Celebrate every victory, no matter how small. Did you sit through a boring wait without reaching for the phone? Hold a phone-free dinner? Close the app when your thumb wanted to keep scrolling? Jot it down and give yourself the credit.

6. **List grateful moments.** Gratitude shifts your mind from what you are missing to what you are gaining, from FOMO to JOMO, the joy of missing out. Each day, write at least one thing you are grateful for, ideally something that could not be photographed or posted. Research by Emmons and McCullough demonstrates that gratitude journaling produces measurable improvements in well-being, sleep quality, and resilience.[24]

7. **Incorporate affirmations.** Write down affirmations that resonate with you. "I am the author of my own attention." "Every day, I grow more present." "I control the device, not the other way around." They act as counterweights to the negative self-talk that tends to drive people back toward the escape hatch.

8. **Note physical changes.** Your body will change as you reclaim your time. Maybe you are sleeping better, your eyes feel less strained, your mind less foggy, your energy steadier. Track these as additional motivation, because they are real and they are yours.

9. **Remember interactions.** Did you have a genuinely present conversation with someone today? Or a moment where the phone pulled you out of one? Document both. They give you insight into your support system and into the places you are still vulnerable.

10. **Plan for the future.** Use your journal to look ahead. What is coming that will test you? A long, dull meeting, a wait at the airport,

a dinner where everyone else will have their phones out, a quiet evening alone with nothing scheduled? Think about how you will handle it, and write the plan down before you are standing in the middle of it.

11. **Reflect on your progress.** Every week, read back over your entries. Recognize the patterns, celebrate the successes, and notice where you might need additional support or a smaller next step.

Remember, your journal is a safe space, free of judgment. It is for you and you alone. You never have to share a word of it, and I strongly advise you never let anyone read it. You do not need to be a perfect writer. What matters is the honesty and the commitment with which you show up to the page. Over time, you will hold a tangible record of your growth, your resilience, and the many small steps you took back toward your own life. Happy journaling.

Measuring Meaningful Outcomes

Beyond daily Screen Time data, your Execute step tracks several broader outcome categories that reflect the actual quality-of-life improvements digital sobriety produces.

Sleep quality: Keep a simple weekly sleep score of 1 to 10 and track your average. Research by the American Academy of Sleep Medicine documents that removing devices from the bedroom and eliminating pre-sleep screen time produces significant sleep quality improvements within two to four weeks.[25] This is one of the earliest and most motivating changes people experience.

Focus and attention: Once a week, attempt a 45-minute deep work or reading session with no phone in the room. Note how far you got before your mind wandered or you felt the pull to check. Over months, your ability to sustain focused attention without a device break should improve meaningfully.

Relationship quality: This is harder to measure but essential to track. Periodically, once a month, ask yourself: Am I more present with the people I care about? Are conversations richer? Are there moments of genuine connection that I was previously absent for? Write about specific instances.

Mood and anxiety: Many people who develop digital sobriety report significant reductions in anxiety, particularly the low-grade ambient anxiety of feeling perpetually behind, perpetually compared, perpetually in need of checking. This is one of the most profound changes, and it takes time, typically 60 to 90 days of consistent practice before people notice a baseline shift.

Self-trust: The most important outcome, and the hardest to quantify. Do you feel that your choices about technology are increasingly yours? Do you feel less pulled, more deliberate? Are there moments where you notice you have chosen not to check your phone and feel satisfied rather than anxious? That growing experience of trusting yourself, of knowing that you can sit with the impulse and choose, is the Flavor of Confidence this method is building.

When the Plan Breaks: Run the Cycle Again

I want to be completely clear about this next part, because it is the single most misunderstood element of the Method, and getting it wrong is what sends people spiraling into shame instead of progress.

Your plan will break. Not might. Will. You will hit a week where the screen-time number climbs back up, where the phone reappears at the dinner table, where you find yourself scrolling at midnight with no memory of having decided to. This is not the Method failing. This is the Method working exactly as designed, because Execute is the step that tests everything you built, and a test is only useful if it can reveal what does not yet hold.

When something in Execute does not go the way you planned, you do not white-knuckle harder and you do not start over from zero. You run the cycle again, deliberately, in this order:

1. **Return to Observe.** Go back to your SOBER Journal and figure out what actually happened. What was the real trigger? Where exactly in the chain did the old behavior take over? Was it a time of day, an emotional state, a specific person, a specific app, or a gap in your routine you had not accounted for? You cannot fix what you have

not honestly seen, so observe it first, with the same radical honesty you brought the first time through.

2. **Move back through Behavior.** Now redesign. Update the behavior chain analysis for the trigger that beat you. Build a new replacement behavior, adjust your environment, change your notification architecture, or add a layer of friction at the exact point the chain broke. The break showed you precisely where the design was weak. Strengthen that spot.
3. **Test again in Execute.** Carry the revised plan back into your daily life and run it. Watch the data. If it holds, you have closed a gap. If it breaks again somewhere new, you already know exactly what to do: back to Observe, through Behavior, and test once more.

This is the loop. Observe, Behavior, Execute, and around again, each pass tighter than the last. It is the Deming cycle of continuous improvement applied to your own attention, and it is why I keep telling you there is no zero in this Method. A slip is not a return to the starting line. It is a data point that tells you where to aim the next cycle. Every place the plan breaks is exactly the place that needs a BCA update, a new behavior chain, or a conversation with your mentor. Resistance is information, not failure, and every time you run the loop you come back more precise than you were before.

The Four Virtues in Execution

The Stoics understood this rhythm of testing and refining long before Deming gave it a name. The Stoic tools you grounded yourself in during the first step, the Dichotomy of Control along with the four cardinal virtues of Temperance, Courage, Justice, and Wisdom, are most needed right here, in the daily friction of living the plan. Let's walk through how each one shows up in Execute.

Dichotomy of Control. As you execute, you will constantly meet the line between what is yours to govern and what is not. You cannot control whether a coworker sends a group text during dinner, whether someone tags you in a post during a device-free hour, or whether an app pushes an update engineered to pull you back in. You can control how you respond to every one of those things. When the plan is working, invest your energy in maintaining it. When it meets an obstacle, spend your energy on the part you can actually change, and refuse to burn it ruminating over what is already done. Seeking validation through the screen is its own version of chasing the uncontrollable, so anchor your sobriety in your own conviction rather than in how the feed responds to you.

Temperance. Temperance is balance, and it means refusing to swing between extreme restriction and total collapse. The goal of this Method was never a device-free life lived in a cave. The goal is a deliberately engaged life in which your devices serve you instead of the reverse. Watch, too, that you do not simply trade one compulsion for another, swapping the phone for overwork, mindless eating, or some other escape. On the hard days, when the old pattern wins and the phone is back in your hand before you noticed you reached for it, temperance says plainly: you are not starting over, you are continuing. There is no zero. There is only forward.

Courage. Courage in execution rarely looks dramatic. It looks like putting the phone in the kitchen at 9:30 PM on the night when everything in you would rather scroll. It looks like being the only person at the table without a phone out. It looks like writing the honest entry in your journal on the day you backslid instead of quietly skipping it. And it takes real courage to admit when you are struggling and to reach out to your mentor or accountability partner rather than hiding it. Reaching out is a sign of strength, not weakness. When your resolve wavers, think back to the harder things you have already survived, and draw on the proof that you can do hard things.

Justice. Of the four cardinal virtues, Justice is the one that reaches beyond you to the people in your life. In Execute it shows up as accountability: following through on what you promised your partner, your children, your friends, and yourself when you said you were working on this. Every kept commitment, however small, is an act of justice toward the people your device habit used to pull you away from, and toward your own integrity. Justice also means fairness to yourself, meeting your own missteps with the same understanding you would offer a friend rather than with punishing self-talk. Because personal accountability sits so close to the heart of this step, it returns later in this chapter for a fuller treatment.

Wisdom. Wisdom is the ongoing refinement of your plan based on what the data actually shows, rather than on what feels comfortable in the moment. It is the pause between the stimulus and your response, the place where deliberate choice lives. Wisdom treats every misstep as a lesson rather than a verdict: you ask what happened, what it taught you, and how you will adjust, and then you fold that answer into your next cycle. It also means knowing your limits and having the sense to put distance between yourself and the people, places, or apps that reliably undo you. The Execute step reveals the limits of your plan with remarkable efficiency, and wisdom is what turns each revealed limit into the starting point for a better design.

Gratitude and Affirmations in Digital Sobriety

I want to pause here and ask you something. How are you doing?

I mean it. Not performatively. Actually. Right now, in this moment of your life, what is going on?

I ask because one of the most powerful practices I have encountered in the Execute step is the simple act of asking someone how they are and then genuinely listening. Not planning your response. Not checking your phone while they answer. Not waiting for them to stop talking so you can talk. Listening.

The phone has, for many people, replaced this capacity. It is easier to comment on someone's post than to sit with them and hear about their actual life. It is easier to send a GIF than to be genuinely curious about what someone else is going through. The restored capacity for real listening, unhurried, undistracted, and fully present, is one of the most profound gains of digital sobriety. It is also, research shows, one of the most powerful builders of the kind of genuine connection that actually meets the belonging need your phone was never quite satisfying.

Gratitude and affirmations are not soft extras; they do real work. Expressing gratitude has been shown to release the same feel-good chemistry, dopamine and serotonin, that your device habit was strip-mining from you, which is precisely the lift you need on the days digital sobriety feels hardest. Gratitude also roots you in the present, where compulsive checking cannot easily reach you, and it shifts your focus from what you might be missing, the engine of FOMO, toward what you are actually gaining. Affirmations do the parallel work on your self-talk. The road away from any compulsion is lined with self-doubt, and deliberately chosen affirmations act as counterweights so the negative voice does not get the last word. Said and re-read consistently, they rebuild the belief in yourself that genuine confidence stands on.

In your daily journal: write one thing you are grateful for that required your physical presence to experience. One thing an algorithm could not have given you.

And write your affirmations. Not generic ones. Write yours specifically. "I am capable of choosing my attention." "I am building a life I do not need to escape from." "I am present." "My Flavor of Confidence grows stronger each day I choose it."

Read these back. They do not work if you write them once and never revisit them.

Mindfulness in the Execute Step

At its core, mindfulness is about being present. It is the practice of anchoring yourself in the current moment, observing your feelings, thoughts, and sensations without judgment.

For device addiction specifically, mindfulness is both the practice and the destination. The goal of the entire Method is, in one sense, an expansion of your capacity to be genuinely present: in your own life, with your own thoughts, with the people around you. Every time you put down the phone and remain with the moment in front of you, that is practicing mindfulness in the most direct way possible.

When the pull toward your device is strong, when you are sitting with boredom or anxiety or the itch of FOMO, try this:

Close your eyes if you can. Take three slow breaths. Focus on the sensation of air entering and leaving. Notice what is happening in your body: where the tension is, what the physical experience of this craving actually feels like. Recognize: This is a feeling. It is not an emergency. It will pass. You do not need to act on every feeling.

The 5-4-3-2-1 technique is particularly useful in device-craving moments: identify five things you can see (without using your phone), four you can touch, three you can hear, two you can smell, one you can taste. This technique grounds you in physical reality, which is exactly where your brain needs to be when the pull toward the virtual is strongest.

Two things make this far more powerful. First, in the moment of craving, gently remind yourself why you began: the presence you want with the people you love, the attention you are reclaiming, the life you are building. Reconnecting to your reason reignites the determination that a craving works to drown out. Second, treat mindfulness as a skill rather than an emergency measure. Set aside a few minutes each day, in the morning or before bed, to practice it when you are calm. Like any skill, it grows stronger with repetition, so that when the strong pull arrives, presence is already a reflex you can reach for rather than something you are scrambling to find.

The Stoic Justice of Personal Accountability

The cardinal virtue of Justice, as it applies to Execute, is ultimately about personal accountability. You have made commitments: to your goals, to your journal, to the people in your life, and to yourself. Following through on those commitments is an act of justice.

Every day you execute your plan, you are contributing to the evidence that you can be trusted, first by yourself, and then by the people around you. The experience of consistently following through on your own commitments, even small ones, is a fundamental component of the self-trust that genuine confidence is built on.

This is why the daily journal is not optional. It is where you record your accountability to yourself in writing, where you see the progress accumulate, and where you confront the days it did not go as planned with honesty rather than avoidance.

The Execute step continues indefinitely. You are always in Execute. As you work through subsequent cycles of S.O.B.E.R., returning to Stoic for deeper reflection, building new behavior chains as your challenges evolve, and setting new goals as old ones are met, Execute is the daily practice that ties everything together.

Now let's move into Restore, where you will address the relational and spiritual dimensions of what digital sobriety is asking you to do.

Notes

Notes

Restore

How are you feeling? More present? More like yourself?

Take a moment with that. Before you do anything else, I want you to grab your SOBER Journal, flip to the third-from-last page, and write at the top: *My Flavor of Confidence.*

Pause. Reflect. Since you began applying the S.O.B.E.R. Method to your device life, what have you noticed? Perhaps your mornings feel less frantic. Perhaps a conversation this week was richer than most conversations have been in years. Perhaps you read something longer than a caption without losing the thread. Perhaps you were at a dinner table and actually tasted the food, actually heard what the person across from you was saying, actually laughed without pulling out your phone to photograph the moment.

Write down at least one thing. It does not need to be dramatic. A quiet shift is still a shift. As you continue and as new moments of genuine presence accumulate, flip to the back of your journal and add them with dates. This is your Flavor of Confidence in the digital realm. These are the moments of intentional living you have begun to recover and build.

Now: I want to tell you something about the Restore step. It is, in my experience, the step people are most tempted to minimize. They have done the philosophical work, the observational work, the behavioral redesign. They are executing their plan. The wins are accumulating. It is tempting to conclude that the hard work is behind them.

It is not. The Restore step addresses something the other four steps cannot: the relational and moral debt accumulated during the years of compulsive device use. Because your phone habit, whatever its scale, has likely cost something beyond your own time and attention. It has cost the people around you. And until you address that honestly, the Flavor of Confidence you are building will have a small hollow place at its center.

"Look well into thyself; there is a source of strength which will always spring up if thou wilt always look." That is Marcus Aurelius, Meditations 7.59. Keep looking. The source is in there, and brighter days are on the horizon.

Now I want to tell you something about the Restore step. It is, in my experience, the step people are most tempted to minimize. They have done the philosophical work, the observational work, the behavioral redesign. They are executing their plan. The wins are accumulating. It is tempting to conclude that the hard work is behind them.

It is not. The Restore step addresses something the other four steps cannot: the relational and moral debt accumulated during the years of compulsive device use. Because your phone habit, whatever its scale, has likely cost something beyond your own time and attention. It has cost the people around you. And until you address that honestly, the Flavor of Confidence you are building will have a small hollow place at its center.

Your Personal Vision Statement

Before we move into the forgiveness work, I want to slow down and spend real time here, because this is the part of Restore I talk about more than almost anything else when I am on a stage or behind a microphone. Done right, it is the single most powerful engine in the entire Method for pulling you forward instead of white-knuckling you forward. Flip back two pages from "My Flavor of Confidence" in your journal and write at the top: Long-Term Goals. We will fill that page shortly. First, we need to know where it is all pointing.

Imagine standing at the helm of your own life, navigating vast open water. To sail anywhere on purpose, you need a fixed point on the horizon to steer by. That fixed point is your personal vision statement. It is not a to-do list, and it is not this quarter's screen-time target. It is a vivid picture of the life you intend to be living in five, ten, even twenty years, written as though you are already standing in it. It spans everything, not just your relationship with your phone: your work, your relationships, your health, your character, the person you want to become, and the legacy you want to leave. Two people can chase nearly identical goals and still write completely different vision statements, because a vision is born out of your own values, your own loves, and your own dreams. It is yours alone.

Here is why it matters so much. On the foggy days, and there will be foggy days, your vision statement is the lighthouse. When a craving hits, when motivation thins, when you cannot quite remember why you put the phone in the kitchen drawer in the first place, you do not have to manufacture willpower out of thin air. You read your vision, and you remember. The pull of a life you are genuinely building will, over time, outcompete the pull of any feed. That is the whole secret of this Method in one sentence: we do not win by hating the phone, we win by loving the life that makes the phone small.

So how do you craft one? Take it in four moves, and give it the better part of an hour in your journal:

1. **Envision the future.** Close your eyes and drop yourself ten or twenty years ahead. Where are you waking up? Who is around you? What does an ordinary morning feel like? What are you proud of? Make it specific and sensory, not abstract.
2. **Reflect on what drives you.** Dig into your values and your loves. What actually matters to you when no one is watching and nothing is being measured? What kind of person do you most want to have become?
3. **Write it down.** Put it on the page in present tense, as if it is already true. It does not have to be polished or final. A rough first draft is exactly right; you will refine it for years.
4. **Revisit and revise.** A map gets updated as the territory reveals itself. Read your vision regularly and adjust it as you grow. The point was never to predict the future with precision. It is to set an intentional direction, so that every step, even the detours, carries you closer to the life you actually want.

Here is an example to write against, not to copy:

In ten years, I wake each morning without reaching for my phone. I sit with my coffee and my journal, and I begin the day as the author of my own attention rather than a consumer of someone else's content. My marriage is close and unhurried because I am actually in it. My children know me as the parent who looked up, who met their eyes when they spoke. My work is the best of my life because I bring whole, unfragmented focus to it, and I am known for the depth of what I make rather than the noise of how often I post. I am strong and healthy because I spend my body's best hours moving through the real world instead of sitting inside a screen. I still use technology, deliberately, as a tool in service of a life I love. And I carry a confidence that needs no external confirmation, grown from the daily practice of choosing presence over compulsion. This is my Flavor of Confidence.

Read that and notice how it makes you feel. I get a little fired up every single time I read one of these, mine included. Now go write yours.

From Vision to Long-Term Goals

A vision statement that only sits on a page is a beautiful daydream. We are going to give it landing gear. The bridge between the horizon and tomorrow morning is your long-term goals, and you build that bridge by working backward from the vision in steadily shortening time frames.

On your Long-Term Goals page, write at least three long-term goals drawn directly out of your vision. For each, think in year-sized horizons: one, three, five, and ten. Some people prefer to set firm goals only out to about three years, because life past that is so volatile. Personally, I set firm goals to five years, then keep looser ten and even twenty year goals beyond that. Do what fits your temperament, but reach further than feels comfortable. The horizon is supposed to stretch you.

For digital sobriety specifically, the cascade out of your vision might look like this:

» **Ten years from now:** I live as a fully present parent, partner, and professional, with technology permanently in its proper place as a tool I command rather than one that commands me.

» **Five years from now:** I have sustained five years of phone-free mornings and phone-free family dinners, and I have built a body of focused work I am genuinely proud of.

» **Three years from now:** My children would describe me, without hesitating, as present. I keep one full screen-free day every week.

» **One year from now:** I have gone twelve consecutive months with no social media after 9:30 PM, kept phone-free meals for a full year, and read twenty-four books I chose on purpose instead of scrolling the equivalent hours away.

Notice that each shorter horizon is not a different dream. It is the same vision, brought one step closer to the ground.

The S.M.A.R.T. Cascade: Making the Impossible Inevitable

Now for the part that genuinely changes lives, and I do not say that lightly. A ten-year vision can feel impossibly far away, so far that it is easy to file it under "someday" and never begin. The trick, the entire trick, is decomposition. You take the impossible-feeling thing and cut it in half, then cut that half in half, and you keep cutting until what is left in front of you is a single step you could take this afternoon. Do that honestly and the impossible does not merely become possible. It becomes nearly inevitable.

You already own the tool for this. Back in the Behavior step you learned to make your goals S.M.A.R.T., that is, Specific, Measurable, Achievable, Relevant, and Time-bound, and then to break them into milestones. We are going to run that exact same tool here, except now we aim it at your whole life and let it cascade down through the time frames.

Here is the cascade, top to bottom:

1. **Start at the ten-year vision.** Pull one thread out of your vision statement. Say it is: I am a genuinely present parent.
2. **Set the ten-year goal, made S.M.A.R.T.** Turn the thread into something measurable and time-bound: "Over the next ten years, my children grow up with a parent who is phone-free during every shared meal, every bedtime, and every school event."
3. **Halve it to five years.** What must be true at the halfway mark for that to stay on track? "Five years from today, phone-free meals and bedtimes are simply how our household runs, unbroken for the

last three years."

4. **Halve it again to one year.** "Twelve months from today, I have held phone-free dinners every night for a full year and phone-free bedtimes five nights a week."
5. **Bring it into this quarter.** "In the next ninety days, I will extend phone-free dinners to seven nights a week and add phone-free bedtimes three nights a week."
6. **Bring it into this month.** "This month, my phone stays in the kitchen drawer from 6:00 PM through the kids' bedtime, five nights a week."
7. **Bring it into this week.** "This week, the phone goes in the drawer at 6:00 PM on Monday, Wednesday, and Friday."
8. **Name today's single step.** "Tonight, the phone goes in the drawer at 6:00 PM, and it stays there until the kids are asleep."

Look at what just happened. "Be a present parent for the next decade" is a goal so large it is almost meaningless on its own. "Tonight, the phone goes in the drawer at 6:00 PM" is something you can absolutely do, today, with no special strength required. And here is the beautiful part: tonight's small, doable thing is not separate from the ten-year vision. It is the ten-year vision, expressed at the only scale any of us ever actually lives at, which is right now. String enough of those nights together and you never have to chase the impossible. You will look up one day and realize you have been living it all along.

A few principles keep the cascade honest:

» **Run every rung through S.M.A.R.T.** If a goal at any level is missing a number or a date, it is not finished yet. Vague goals are where good intentions quietly go to die.

» **Let the far horizons stay a little loose.** Your ten and twenty year goals are direction, not contract, and you should expect to revise them. It is the near rungs, this quarter, this month, this week, and today, that need to be sharp.

» **Review on a rhythm.** Each week, look at the near rungs and check them off or adjust them. Each pass through the Method, lift your eyes back to the vision and make sure the ladder is still leaning against the right wall.

» **Adjust the plan, never the destination.** If you fall behind, you do not lower the vision. You re-cut the next chunk into a smaller, more achievable step, and you keep climbing.

This is how ordinary people build extraordinary lives. Not in one heroic leap, but by turning a distant vision into a sequence of steps small enough to actually take, and then taking the next one. Write your vision. Build your ladder all the way down to tonight. Then go take tonight's step.

Read your vision statement regularly, especially when the Method gets hard. It is your guiding star and your grounding anchor, both at once. Return to it, and let it pull you home.

Self-Forgiveness

The Restore phase begins with you, and the first act it asks is self-forgiveness.

You have made mistakes. You will make more. You may even slip back unexpectedly into the very patterns you are working to break. It happened to me early in my own sobriety, and forgiving myself was one of the hardest things I ever had to learn. I was so ashamed for slipping that the shame itself nearly drove me back under. Learning to forgive myself turned out to be the missing piece, one of the most important parts of staying well at all. So we start here, with you.

This is harder than it sounds for device addiction, because the damage is often diffuse and incremental in a way that makes it easy to minimize. You did not do anything illegal. You did not crash a car or lose a job (probably). The harm accrued in smaller, quieter ways: in a child's recital you watched through a screen instead of your eyes, in a conversation you half-participated in while scrolling, in years of sleep you degraded without fully realizing it, in hours, thousands of them, you gave away to an algorithm while your actual life was happening elsewhere.

The quietness of this harm does not make it less real. It makes self-forgiveness both more necessary and more subtle.

Self-forgiveness in this context is not about declaring that the time you lost was fine. It is about acknowledging it clearly, understanding the context in which it happened, for you were captured by a system engineered by some of the world's best engineers to capture you and you were not unusual in being captured, and deciding that you are not going to carry the weight of it forward as shame. Shame is not a useful engine for change. It creates the very anxiety it claims to address, and that anxiety tends to drive people back toward the exact escape hatch they are trying to close.

The alternative to shame is honest accountability. Not self-flagellation. Not denial. Seeing what happened clearly, making specific amends where you can, and committing forward.

Your journal is the most powerful tool you have for this. Putting pen to paper makes you confront what happened head-on, and the act of writing it out provides a genuine emotional release, a way to set the weight down on the page instead of letting it sit and fester inside you. It also organizes the chaos, because regret and guilt tend to swirl, and a written narrative turns fragmented, looping memories into something you can actually look at and address. Months from now, on a day when self-forgiveness feels out of reach, you will be able to read back and see exactly how far you have come. Be genuine when you write, even when it is uncomfortable, and watch your language as you go: the purpose here is to heal, not to hand yourself one more reason to feel small.

In your SOBER Journal, write about three specific moments in which your device use cost someone else your presence. Be specific: name the person, the situation, what you did instead of being present. Then write about what you wish you had done differently. Then write the following, or something like it in your own words: I am accountable for this. I cannot change it. I have learned from it. I will be different going forward. I forgive myself for the person I was before I knew better.

Read it aloud if you can.

Forgiving Others, and What Forgiveness Really Is

Self-forgiveness is where Restore begins, but it is not where it ends. This step asks three distinct acts of forgiveness of you, and the Method asks that you complete all three toward at least one person before you cycle back to Stoic: forgive someone who wronged you, seek forgiveness from someone you wronged, and make amends. We have done the first kind, turned inward. Before we turn to seeking forgiveness, spend a moment on the act people most often skip entirely: forgiving someone else.

You may wonder what forgiving other people has to do with your phone. The connection is direct. Resentment is a weight. Carrying anger or bitterness toward someone is a low, constant background ache, and background aches are exactly the emotional states that send your hand toward the phone for relief. Every grudge you hold is, quietly, another trigger. Setting those weights down is not only good for your soul, it removes one more reason you reach for the escape hatch.

So let me demystify what forgiveness actually is, because it is badly misunderstood.

Forgiveness starts and ends with you. It is an inner decision to release the bitterness you carry toward someone who harmed you. The common question, "do they even deserve my forgiveness?" is the wrong question, because forgiveness is not about them. It is about you. Holding onto hurt is like carrying a heavy pack everywhere you go. Forgiveness is the act of setting it down. You are choosing your own peace over a festering wound.

A few things forgiveness is not:

» **It is not an event. It is a process.** There will be smooth stretches, rough patches, and some backtracking. Take your time and heal at your own pace.

» **It is not weakness.** It is one of the strongest and most mature things a person can do: choosing understanding over animosity, and forward motion over being stuck in the past.

» **It is not forgetting.** "Forgive and forget" is misleading. You can remember exactly what happened and simply stop letting the memory hold emotional power over you. When old feelings resurface, and they will, you just reaffirm your choice to forgive.

» **It is not excusing or condoning.** You can see an act as genuinely wrong and still free yourself from its grip.

» **It is unconditional.** Receiving an apology can help, but your forgiveness should never be held hostage to someone else's remorse. They may never apologize. Forgive anyway, for your own sake.

And forgiveness is not the same as reconciliation. This distinction matters enormously, so hold onto it. Forgiveness is something you do alone, for your own peace. Reconciliation involves two parties and the rebuilding of a relationship. It requires mutual understanding, honest communication, and usually some agreement about how to move forward, and it works only when both people recognize the harm and are willing to rebuild trust. You can fully forgive someone and still decide, wisely, to keep your distance. Whether to reconcile depends on the depth of the harm, the other person's willingness to make amends, and your own boundaries. You always have the autonomy to forgive. Reconciliation is a separate, collaborative choice, and you are never obligated to it.

In your SOBER Journal, name one person you are carrying resentment toward. Write what happened, and then write your way toward setting it down, not because they earned it, but because you are tired of carrying it. Keep the decision to forgive separate from any decision about whether to rebuild the relationship. The first is yours alone. The second is not.

Seeking Forgiveness and Making Amends

Forgiving yourself is the foundation. Seeking forgiveness from the specific people your device habits harmed is the structure you build on top of it.

This is the part of the Method that surprises people most. "Apologize to someone for being on my phone too much?" Yes. And I want to explain why it matters far more than it sounds.

The people who have experienced the cost of your device habits, among them a partner who felt perpetually competing with your screen for attention, a parent whose calls you consistently let go to voicemail, a child whose moments you half-witnessed through the distraction of a feed, carry something from that experience. They may not have named it. They may not even be fully conscious of it. But the experience of feeling less important to someone than their phone is real, and it leaves a residue.

Seeking forgiveness is not about re-litigating the past. It is about acknowledging that it happened, that the other person's experience was real and valid, and that you see it now.

I have had remarkable experiences with this step. I have been received with immediate warmth and relief, by people who had apparently been waiting for acknowledgment without expecting it. I have been received with anger, which I sat with calmly because I recognized that I had earned it. In one case I was met with an unexpected apology from the other person, who said sorry for not having been in a good place themselves back then. We just sat, listened, and forgave each other, and we parted with a long hug, better than we had been in years. I am 38 for 41 in the formal forgiveness conversations I have initiated for my device work, and 117 for 129 across my whole sobriety. The ones that did not land simply did not; those people were not ready, and I made my peace with that. The point was never a perfect record. The point was that I showed up and offered it.

Choosing the person. I generally figure out who my person is going to be when I return to Stoic, and I recommend you do the same: pick them within the first few weeks of your cycle. Sometimes it has taken me weeks just to land on the right name from my past, and that is fine, because settling on it early gives you a month or two to sit with what actually happened between you. That stretch of reflection is not wasted time. It is how you work all the way through your own pain so you can walk into the conversation prepared for anything. Make notes in your journal. Be thoughtful about how you want to show up when you meet.

Making contact. Reach out and meet in person if you can, somewhere public. Grab a coffee. People are generally less likely to raise their voice in a coffee shop, and a public table keeps things calm. How do you even open that door? It is easier than you think. Picture someone who hurt you years ago reaching out and saying, "Hey, I feel bad about what happened back then.

Could we meet for a coffee? I would really like to apologize in person." You would probably say yes. Most people do, and I have been genuinely shocked at how willing they are to give me twenty minutes. When you get there, buy their coffee. Manners matter.

Every so often someone will push back and say they would rather not meet in person. Do not force it. Reply with something like, "I understand. I would be skeptical of me, too. Could we talk for a few minutes on the phone instead? I really just want to listen to what you have to say." More than once, that gentle reply has turned into "Okay, I'll grab a coffee with you" after all. Take that as the win it is.

The conversation itself. Listen. Let me say it again, because it is the whole game: listen before you reply. You have your coffee, you have exchanged pleasantries, and now it is on you to begin. In your own words, something like this:

> *"Let me start by thanking you for taking the time to meet me in person. I have been on a journey to be more present for the people in my life, and part of that is apologizing to the people I was not present for. There were too many times I was with you but not really with you, looking at my phone instead of giving you my actual attention. You deserved better than that, and I am genuinely sorry. Is there anything I can do to make it right?"*

Then stop talking. Anything could happen next, so listen, reflect, pause, and only then respond. Do not interrupt, for any reason. Every rule of conversational courtesy you have ever heard, employ it here. You are not defending yourself and you are not explaining yourself. Your steady, attentive presence, sitting with whatever they have to say, is the apology.

When it gets hard. Sometimes, despite your best intentions, the conversation turns. Be prepared for it. You might even hear something like, "I hate you, lose my number." If that happens, respond calmly and only once, and refuse to be pulled into a long back-and-forth. Some people will not be ready to understand or forgive, often because of what they are carrying themselves, and that is not yours to fix. Give yourself real credit for making the effort, and protect your own journey, even if that means blocking the number afterward. More than once, someone has come back to me years later to apologize for not having been ready the first time.

If the conversation stays in the room but gets heated, a few things help:

» **Stay composed.** Take a breath and a beat before you answer. Respond rationally, not reactively. Meeting heat with heat only raises the temperature.

» **Listen actively, with nothing queued up to say.** People often turn confrontational precisely because they feel unheard. Look at them. Nod. Offer small affirmations like "I understand" or "I see where you are coming from." Being truly heard lowers the temperature more than any argument can.

» **Use "I" statements.** Instead of "You're not hearing me," try "I feel unheard when my point gets lost." It keeps blame out of the exchange. Then wait, calmly, until they hand you the floor. They will.

» **Ask open-ended questions.** "Can you help me understand your perspective?" or "What makes you feel that way?" shows you actually want to understand rather than win.

» **Set a boundary if you need one.** "I value this conversation, but I think we both need to keep it calm and respectful. Could we take a short break and pick it up again?" It is okay to pause and revisit another time.

Afterward, reflect on what happened. What set it off? Were there early signs? The aim of every one of these conversations is mutual understanding, never victory.

Do not reach for reconciliation. This conversation is about acknowledgment, not about rebuilding the relationship, and you should not steer it there. If they float the idea of getting together sometime, keep it light and noncommittal: "That would be nice. Work has me pretty buried right now. Why don't you reach out in a couple of months and we'll see?" Notice there is nothing definite in that. In most cases you will not hear from them again, and, strange as it sounds, that is usually a good thing. As you keep growing, you naturally outgrow some of the people from the chapter you are leaving behind. I have a saying for it: "If you are the only present person in the room, you're in the wrong room." Look for the people who pull you forward, who challenge you to keep your attention where you want it and to keep becoming who you are working to be. You will form new and stronger bonds with them.

Rehearse it first. Do not improvise this. Role-play the whole thing with your accountability partner before you sit down for real, and have them throw the hard responses at you: anger, dismissal, sudden tears, stony silence. Sobriety of any kind is really just practicing, every single day, the things you do not want to be caught unprepared for. A conversation this important is worth practicing too.

The Restorative Power of Volunteering

Device addiction is, among other things, a narcissism accelerator. Not in the pejorative sense, but in the literal sense that social media and notification-driven engagement consistently orient your attention toward yourself: your image, your metrics, your online reputation, your emotional reactions to what others are doing. Hours of daily self-focused activity gradually narrows the psychological aperture through which you see the world.

Volunteering reverses this. When you give your time and attention to a cause larger than yourself, such as a community garden, a literacy program, a food bank, a shelter, or a youth organization, the psychological effect is well-documented and significant. A 2013 meta-analysis by Jenkinson and colleagues, published in *BMC Public Health*, found robust associations between volunteering and improved well-being, sense of purpose, reduced depression, and increased social connection.[26] The effect size was comparable to other major well-being interventions.

For digital sobriety specifically, volunteering offers several additional benefits:

» **Replaces scrolling time with embodied, present-tense activity.** You cannot scroll while carrying boxes at a food bank or reading to elderly residents at a care facility.

- **Provides genuine social connection.** The sense of belonging volunteering creates is built on shared real-world experience, not digital performance.
- **Rebuilds sense of purpose.** Many compulsive device users describe a background feeling of meaninglessness, a sense that the time they spend online does not actually add up to anything that matters. Volunteering is antidote to that.
- **Generates genuine self-respect.** The kind of esteem you cannot get from follower counts.

So how do you find the right fit? Start with what you actually care about. Do you love animals, the outdoors, kids, the elderly, books? Let that point you. Then look close to home first, because community centers, schools, hospitals, libraries, food banks, and places of worship almost always need hands. Online platforms like VolunteerMatch and Idealist let you plug in your location and surface dozens of options nearby, and a local volunteer fair is a great way to meet several organizations at once. Ask the people in your life, too; word of mouth turns up more than you would expect. If you are new to this, start with a short-term commitment so you can feel out whether it fits before going deeper, and do not be shy about asking plain questions up front about the work, the time it takes, and the support you will have. Then, once you have chosen, dive in with your whole heart. Volunteering is not just about giving your time; it is about giving a piece of yourself, and the more of yourself you bring, the more it gives back.

I had never volunteered until a judge required it. I walked in like I was the CEO of the whole operation, found myself organizing the least motivated people in the room, and gave it my usual full-tilt energy. The event organizer was impressed enough to send a letter to the judge on my behalf, without me asking, and almost nobody does something like that anymore. I have been volunteering somewhere ever since. Find something you care about, bring your full energy to it, and watch what it does to your relationship with the whole world, including your phone.

Mentoring Others in Digital Sobriety

Before you consider taking on a mentee in digital sobriety, reflect on a few questions:

» Have you maintained meaningful, measurable progress in your own digital sobriety for at least six months?

» Are you emotionally stable enough in your own practice to support someone else without destabilizing yourself?

» Do you have genuine familiarity with the S.O.B.E.R. Method, including the ability to guide someone through BCA construction and goal-setting?

» Do you have the time and sincere willingness to be consistently available?

» Are your reasons for mentoring genuinely about the other person?

If the answers are yes, then mentoring is one of the most powerful reinforcements of your own practice available to you. Teaching the Method consolidates your own understanding of it. Witnessing someone else navigate the early stages reactivates your own motivation. Being the person someone calls when their resolve is wavering gives you a reason to maintain yours.

When you mentor:

1. **Listen more than you speak.** Your job is not to deliver information. It is to help your mentee see themselves clearly. Often what they need most is simply to be heard without judgment.
2. **Be available consistently.** The early stages of digital sobriety come in peaks of intensity: the first week without social media notifications, the first dinner where everyone else has their phone out, the first slip back into old patterns. Your mentee needs to know you will pick up when they call. Consistency is how trust gets built.
3. **Share your experience selectively.** Your story is one data point, not the point. Being open about your own struggles and wins makes their road feel less lonely, but never let the relationship become about you.
4. **Guide, do not dictate.** Offer suggestions grounded in the

Method and your own experience, then respect their autonomy. People hold onto the decisions they reach for themselves far longer than the ones handed to them.

5. **Keep what they tell you confidential.** Trust is the bedrock of this relationship. What your mentee shares stays with you, unless someone is at genuine risk of harm.
6. **Set and respect boundaries.** Your own sobriety comes first, always. Be clear about when and how they can reach you, and if the relationship begins to cost more than you can give without endangering your own practice, have the honest conversation.
7. **Stay rooted in your own journey.** You are still working the Method yourself. Your continued practice is the foundation everything you offer them stands on.
8. **Be honest about your limits, and never shame.** No one has every answer, so when you are out of your depth, say so and point them toward more help. And when they slip, and they will, help them understand what happened, update their BCA, and keep moving. Disappointment and comparison have no place here.

As I said at the end of the Restore step in every version of this Method: helping others heals you. The evidence here is not only anecdotal but biochemical; acts of genuine service increase serotonin and oxytocin production in both the giver and receiver.[27] The loop closes beautifully: you help others find digital sobriety, and in doing so you reinforce your own.

You have now done the full work of Restore: you have forgiven yourself, you have forgiven another, you have sought forgiveness from someone you harmed, and you have made amends. That completes one full pass of the S.O.B.E.R. Method. Do not skip any of those acts on your way out; they are the requirement that closes the loop.

So now it is time to return to Stoic and run the cycle again. Here is what I want you to watch for: the view of yourself starts to change. You will not just feel it, you will see it, measured in your metrics and written in your own hand across your journal. That is not because your earlier passes were wrong. It is because you have grown. This patient, continual inventory of who you are, run cycle after cycle, is what carries you toward the best version of yourself. It is not a quick path. It is an investment worth every bit of your time and energy.

You will start to notice other people treating you with more respect and more love. And with any luck, like me, you will arrive somewhere I once believed was closed to me for good: you will find that you are finally capable of loving yourself, and from there, the love you give to everyone else feels different, fuller, real. Words do not quite reach the change you feel once that happens.

I am certain you are already feeling better about yourself than the day you started. Your Flavor of Confidence is growing stronger every single day.

Go get 'em!

NOTES

Notes

When You Slip: The Next Ten Minutes

You have now walked the full S.O.B.E.R. cycle, from Stoic through Restore. Before we step back to look at what makes this Method different, I want to hand you one thing to keep close, because you will need it.

You are going to slip. I have said this several times throughout the Method, and I will say it once more here, on its own page, because the moment you slip is the moment most people quietly abandon the whole effort. They lose two hours to the feed, feel the wave of shame, decide they have failed, and use that supposed failure as permission to stop trying. I do not want that to happen to you, so here is something concrete to do in the exact moment it happens.

When you surface from a scroll you did not choose, when you realize the phone is in your hand and you do not remember picking it up, when an hour has gone somewhere and you cannot say where, do this. It takes about ten minutes, and it turns a slip into the most useful kind of data.

1. **Stop, and set the phone down.** Physically put it out of reach. Not in your pocket. Down, and away.
2. **Name it without shame.** Say it plainly, out loud if you can: "I slipped." Not "I failed." Not "I always do this." Just the fact, without the story you want to pile on top of it.
3. **Take one slow breath.** One. In, and out, slower than feels natural. You are interrupting the spiral, not solving your life.
4. **Open your journal and write the trigger.** What were you doing in the minute before you reached for the phone? What were you feeling? This is a behavior chain in miniature, and it is worth more than the hour you lost.
5. **Do one small physical thing.** Stand up, stretch, fill a glass of water, step outside, anything that puts you back in your body and out of your head.
6. **Return at the next scheduled check.** You do not start the day over. You do not start the Method over. You simply rejoin your plan at the next point it asks something of you.
7. **Let it go.** You have learned what you needed to learn from it.

Carrying it any further only feeds the anxiety that drove the slip in the first place.

That is the whole protocol. There is no zero, and there is nothing to start over. Every slip you walk through this way is not a step backward. It is a behavior chain you now understand better than you did an hour ago.

The S.O.B.E.R. Difference

As you put the S.O.B.E.R. Method into practice for digital sobriety, you may notice several distinct differences from other approaches to screen-time management. I am confident these differences give you a significant advantage.

It is not religion-based. Let me be more honest with you here than I was in my first book, where I simply told you I was a Christian and left it at that. The fuller, rawer truth is this. When I created the S.O.B.E.R. Method, I was a nihilistic atheist. I believed in nothing beyond the matter in front of me, and I built this Method on Stoic philosophy in part because it asked nothing of me on the question of God. What I did not expect was what the daily practice of Stoicism would do to me over the years.

Sitting with my own mind every morning, examining my actions, reckoning honestly with what I could and could not control, slowly pried back open questions I had long ago decided were closed. The more I practiced, the more I found myself asking about the greater meaning of things, until, to my own genuine surprise, I reconciled with the Catholic faith I had been baptized into at birth. So I will tell you plainly: I am Catholic today. And I will tell you just as plainly that you will find none of that in this Method, by design. The S.O.B.E.R. Method is built on Stoic philosophy, not on any religion, because device addiction reaches people of every faith and of no faith, all over the world. The framework is open to everyone, compatible with every major tradition, and it asks for no belief beyond a willingness to look honestly at your own mind. Where that honest looking eventually leads you is yours alone, as it was mine.

It measures across multiple dimensions. Most screen-time advice asks a single question: Are you on your phone less? Yes or no? The S.O.B.E.R. Method tracks sleep quality, relational presence, attention span, mood, self-trust, and long-term goal progress. This multi-dimensional measurement gives you a much richer picture of what is actually improving, and ensures that the changes you are making translate into genuine quality of life rather than just smaller screen-time numbers.

It is not a label. The S.O.B.E.R. Method explicitly discourages statements like "I am a phone addict" or "I struggle with technology." Here is why that matters. There is a well-documented principle in psychology called the self-concept threat: when we identify as a problem, rather than as a person experiencing a problem, our brain treats attempts to change as threats to our identity, and unconsciously works against them. Constantly affirming that you are an addict makes you one, in a neurological sense, because it reinforces the neural identity structures associated with the addictive behavior. You are not a phone addict. You are someone who developed compulsive device habits in an environment engineered to produce exactly those habits, and you are the person actively changing your relationship with technology. Own your progress, not your diagnosis.

It is fully personalized. Unlike one-size-fits-all digital detox programs, the S.O.B.E.R. Method begins with your specific triggers, your specific BCAs, and your specific goals. Two people using this method side-by-side will build completely different behavior plans because their device relationships are completely different. Your Flavor of Confidence in the digital realm is uniquely yours.

It addresses the root. Screen-time apps and digital detox challenges target the surface behavior. The S.O.B.E.R. Method targets the need that the surface behavior is meeting. This is why sustainable change is possible with the Method and difficult without it. You cannot discipline your way out of a need. You can only find better ways to meet it.

It does not shame. If you are struggling or have had a setback, the S.O.B.E.R. Method does not require you to start over, declare failure, or carry the weight of relapse as evidence that something is fundamentally wrong with you. The method is a continuous improvement cycle. Setbacks are data. Every time you return to the cycle after a slip, you are not starting over; you are advancing. Each pass through the cycle builds on the ones before it.

It builds something positive. Most digital wellness programs frame their entire project in terms of what you give up: less phone time, fewer apps, less social media. The S.O.B.E.R. Method frames its project in terms of what you build: your unique Flavor of Confidence, your deepened relationships, your recovered attention, your richer inner life. The goal is not a diminished existence; it is a more fully inhabited one.

Finally: you are not powerless over your device. That belief, that the phone has you and you cannot change it, is itself a habit of thinking, not a fact. Hundreds of millions of people have learned to develop and maintain intentional relationships with their technology, across all ages and starting points. You have the capacity to do exactly this. The method is a framework for developing that capacity systematically and measuring your progress along the way.

Notes

Device Addiction is Real

A couple of months after my first book was released, I received that letter I described in the Preface, from a reader who had used the S.O.B.E.R. Method to break their device addiction. I told the story there. What I did not tell was the full arc of what came after.

That letter opened a door I have spent three years walking through to write this book. I began actively working with volunteers, people who described their relationship with their devices in terms that were clinically indistinguishable from the language of addiction: the compulsion, the repeated failed attempts to stop, the shame, the progressive displacement of other life activities, the relief that turned to regret. Two of my participants exhibited what appeared to be genuine physical withdrawal symptoms, shaking hands, elevated heart rate, and acute anxiety, when separated from their devices for extended periods. That finding surprised me. I do not think it will surprise the researchers.

The science of device addiction has developed with remarkable speed over the past decade, and its conclusions are not comforting. Let me share what the evidence actually shows, because the why behind the Stoic foundation, the Observe tracking, and the Behavior interventions in this book is grounded in biology and psychology, not opinion.

How It Works: The Neurological Architecture of Device Addiction

The same neurological system that drives substance addiction drives compulsive device use: the mesolimbic dopamine pathway.

Dopamine is a neurotransmitter primarily associated not with pleasure itself but with the *anticipation* of reward. When your phone buzzes, your brain releases dopamine before you even know what the notification is about. The uncertainty is, neurologically, the point. The psychologist B.F. Skinner demonstrated decades ago that *variable ratio reinforcement*, meaning reward that comes unpredictably, produces the most persistent and compulsive behavior of any reward schedule.[28] Slot machines operate on this principle.

So does your social media feed. The next scroll might bring something interesting or nothing at all; the unpredictability is precisely what makes it so hard to stop.

Over time, repeated dopamine spikes from device notifications raise the brain's baseline dopamine threshold. The technical term is *dopaminergic downregulation*: your brain reduces its sensitivity to dopamine to compensate for the excess stimulation. The practical result is that everyday life, including conversation, nature, quiet reading, and the texture of your own thoughts, begins to seem less satisfying by comparison with the device. This is not a character flaw. It is neurological adaptation to an abnormal stimulus environment.

Dr. Anna Lembke, Medical Director of Addiction Medicine at Stanford and author of *Dopamine Nation*, describes this dynamic as the "pain-pleasure seesaw": each dopamine spike from the device is followed by a compensatory dip into what she calls the "pain side," namely anxiety, restlessness, and dissatisfaction, which creates the drive to reach for the device again.[29] The cycle is self-reinforcing and, over time, self-escalating.

The Scale of the Problem

The data on smartphone and device use in recent years is striking in its consistency.

A 2023 Reviews.org survey found that nearly 57% of U.S. adults described themselves as "addicted" to their phones, with Gen Z the generation most likely to feel anxious after losing their phone and the quickest to check a notification.[1] A 2024 Gallup study reported that American teenagers spend an average of 4.8 hours per day on social media alone, not counting other device use.[30] Global average smartphone screen time now exceeds four hours per day for adults across most high-income nations.[7]

Jonathan Haidt's 2024 book *The Anxious Generation* synthesizes a decade of research linking the smartphone and social media revolution, which he dates to approximately 2012, the year both Facebook and Instagram became widely mobile-accessible, with a dramatic and well-documented rise in anxiety, depression, self-harm, and loneliness among adolescents, particularly girls.[4] The data Haidt compiles across multiple large-scale longitudinal studies is sobering: rates of adolescent depression, anxiety, and loneliness began rising sharply in the early 2010s and have continued upward across multiple nations with different social policies, different cultures, and different economic conditions.[31] The one common variable is the smartphone.

Haidt's analysis extends to adults, though the effects are less dramatic, likely because adult brains, being more fully developed, are more resilient to the specific harms of social comparison and algorithmic engagement, and because adults generally have more established identities and social structures to return to. But the adult data is not reassuring either.

Mental Health Outcomes

A comprehensive 2024 meta-analysis by Lam and colleagues, published in the *Journal of Behavioral Addictions*, analyzed 94 studies covering over 680,000 participants across multiple countries.[2] Their findings: problematic smartphone use was significantly associated with elevated symptoms of anxiety (odds ratio 2.14), depression (OR 2.43), loneliness (OR 1.87), and impaired sleep quality (OR 2.08). Effect sizes were consistent across age groups, genders, and cultural contexts.

A separate 2024 systematic review by Vahedi and Saiphoo examined 53 peer-reviewed studies on social media use and well-being.[32] Their conclusion: passive social media use (scrolling without posting or interacting) was the highest-risk activity, associated with upward social comparison, FOMO, and depressive symptoms. Active use (direct communication, intentional content creation) carried significantly lower risks. This distinction is practically important: the harm is concentrated in the mindless scrolling that characterizes most people's device use, not in intentional, bounded engagement.

These findings have a direct implication for the S.O.B.E.R. Method. The goal is not eliminating technology. The goal is transforming passive, compulsive engagement into active, intentional engagement, which the research suggests carries genuinely different mental health implications.

The Attention Fragmentation Crisis

One of the less discussed but most pervasive effects of smartphone overuse is what researchers call *attentional fragmentation*: the progressive deterioration of the capacity for sustained, focused attention.

Gloria Mark, Professor of Informatics at UC Irvine, has studied workplace attention and digital interruption for over two decades.[15] Her research documents a troubling trend: in 2004, the average time a worker maintained focus on a single task before switching was 2.5 minutes. By 2012, it had fallen to 75 seconds. By 2020, it was 47 seconds. The primary driver of this decline is not overwork or cognitive limitation; it is the constant interruption architecture of connected devices.

The 23-minute recovery time Mark documents (the average time to fully return to a task after a digital interruption) means that in a workday punctuated by 65 to 80 notifications, it is virtually impossible to ever achieve sustained deep focus. The implications extend beyond productivity into quality of life: research consistently links the capacity for sustained attention with measures of happiness, creativity, and a sense of meaning. The philosopher William James wrote in 1890 that the ability to voluntarily return the wandering attention is "the very root of judgment, character, and will."[33] Smartphones are quietly eroding that capacity in tens of millions of people, every day.

There is good news here: this is not permanent. Neuroscience research on neuroplasticity demonstrates that the attentional capacity the phone has progressively diminished can be rebuilt through deliberate practice. The practices in the Behavior and Execute steps of this book, including extended phone-free periods, daily reading, and device-free deep work, are precisely the activities that allow the brain's attentional systems to recover. The process takes months, not days. But it works.

Sleep Disruption

The relationship between smartphone use and sleep quality is among the most consistently replicated findings in the research literature.

Blue-spectrum light emitted by smartphone screens suppresses melatonin production for up to 90 minutes after exposure, delaying sleep onset.[12] Social and news content triggers cognitive and emotional arousal, including anxiety, excitement, curiosity, and envy, that further delays the transition to sleep. The anticipatory check for notifications during the night interrupts sleep architecture.

A 2024 large-scale study by Exelmans and Van den Bulck, published in *Sleep Medicine*, found that bedtime smartphone use was associated with significantly later sleep onset, shorter total sleep duration, and lower perceived sleep quality in a sample of over 14,000 adults.[14] The effects were dose-dependent: more bedtime use predicted worse sleep outcomes.

Cumulative sleep debt has consequences far beyond tiredness. Insufficient sleep elevates cortisol (the primary stress hormone), impairs prefrontal cortex function (reducing capacity for impulse control and decision-making), and is associated with increased emotional reactivity and vulnerability to anxiety and depression. In a particularly cruel irony: sleep deprivation increases the compulsive pull toward device use, because the fatigued prefrontal cortex that is responsible for impulse control is exactly the brain region most impaired by poor sleep.

Moving your phone charger out of your bedroom is not a small convenience change. It is one of the highest-impact single actions available to you in this entire book.

The Relationship Cost

The relational cost of compulsive device use is real, documented, and, I want to say this carefully but directly, one of the most serious long-term harms of the problem.

The phenomenon researchers call "partner phubbing" (ignoring your romantic partner in favor of your phone) has been studied in multiple countries. A 2023 study by Roberts and David in *Psychology of Popular Media* found that partner phubbing was significantly associated with lower relationship satisfaction, higher conflict frequency, and reduced trust, with effects mediated by the phubbed partner's increased feelings of exclusion and reduced sense of belonging within the relationship.[9]

The parent-child version of this dynamic carries particular gravity. A 2022 study by Yildiz published in *Current Psychology* found that parental smartphone use during child-directed activities was associated with children's reduced secure attachment, increased emotional dysregulation, and higher rates of externalizing behavior problems.[10] Children experience parental phone use as rejection at the level of attachment security, even if the parent is physically present.

I want to say this clearly: I am not writing to produce guilt. Guilt is counterproductive; it tends to produce the avoidance and self-medication it claims to oppose. I am writing to give you an accurate picture of the stakes, because people make better decisions when they understand what is actually at risk. Your relationships, with your partner, your children, your parents, your friends, and your colleagues, are your most valuable assets. They are also the assets most quietly damaged by the habit of absent presence.

The Restore step asks you to make at least one amends. If you have children, the amend you owe them does not require a conversation; it requires showing up, full-facedly present, in the moments that still lie ahead. That is enough. Start there.

What the Interventions Actually Achieve

The research on digital wellness interventions is not merely supportive of trying; it is specifically supportive of the types of interventions embedded in the S.O.B.E.R. Method.

Grayscale mode: As noted in the Behavior chapter, a 2024 RCT published in *PLOS ONE* found that enabling grayscale screen settings reduced daily screen time by an average of 24 minutes per day, with participants reporting improved perceived control and lower notification-related stress.[22] A separate 2024 study by Kim and colleagues replicated these findings and found additional effects on mood stability.[34]

Notification reduction: A randomized controlled trial by Kushlev and Dunn, published in *Computers in Human Behavior*, found that participants who reduced notification frequency reported significantly lower stress, higher concentration, and improved mood.[21] The effect did not require abstaining from phone use; simply checking phones on their own schedule rather than at the phone's prompting was sufficient to produce benefits.

Digital detox periods: A 2024 systematic review by Stieger and Lewetz, examining 24 studies on digital detox interventions of varying lengths, found consistent evidence for improved well-being, reduced anxiety, improved sleep, and stronger sense of social connection following intentional device-free periods.[35] Effects were largest for interventions of one week or longer, but even 24-hour breaks showed measurable benefits.

Behavior-based interventions: A 2025 systematic review by Van Deursen and colleagues, examining 31 studies on behavioral self-regulation interventions for problematic smartphone use, found that structured behavior change programs, including behavior chain analysis, goal-setting with specific metrics, and social accountability, produced the largest and most durable reductions in problematic use, outperforming both technology-based restrictions and generic awareness-raising approaches.[36]

This last finding is worth pausing on. The S.O.B.E.R. Method is, at its core, exactly the type of structured behavior change program this research identifies as most effective: it includes behavior chain analysis (Observe step), specific goal-setting with metrics (Behavior and Execute steps), and social accountability (network building, mentoring, forgiveness conversations in Restore). The convergence between ancient Stoic practice and modern behavioral science is not an accident; they are both grounded in accurate observations about how human minds actually change.

Why the Stoic Foundation Matters

Jonathan Haidt identifies the smartphone as a "great rewiring of childhood," but he also identifies the solution, at least in part, as the restoration of something essentially Stoic: unstructured time, internal orientation, the capacity to manage one's own emotions without external rescue.[4] His recommendations in *The Anxious Generation* align with the core Stoic practices embedded in this book: reduced social media use, increased real-world interaction, restored capacity for independent thinking, and tolerance for discomfort rather than immediate escape.

The Stoic tradition understood something that modern behavioral science has independently confirmed: genuine confidence, the kind Marcus Aurelius cultivated through private discipline while governing an empire, is built from the inside out, not the outside in. It is built through the repeated experience of facing difficulty and finding that you are adequate to it. Every time you put the phone down and stay present with your own experience, you are building that confidence. Every time you resist the pull of the feed and remain with a moment that does not need to be documented or shared, you are practicing it.

Your Flavor of Confidence in the digital age is not a number on a screen-time dashboard. It is the lived experience of being present in your own life, trusted by yourself, available to the people who matter to you. It is available to you. The Method is the path.

Notes

NOTES

Continuous Improvement

I remember the very early days of my lasting sobriety, when I was still not sure I could find the strength to put together one more sober day, let alone a whole life. People ask me about that stretch all the time now, on stage and in interviews. What was different that time? Why, after everything, did it finally hold? I have thought about it a great deal, and the honest answer runs straight through the people and the practice that caught me.

In October 2015, while I was in rehab, a man came in to share his story, and it never left me. He had grown up in a broken home, dropped out of high school, and ended up on the street in the grip of a brutal crack addiction, stealing and selling to survive until survival was the only thing left. Somehow, against every odd, he found his way out. He spoke about meeting the woman who became his wife, about their two sons in high school now, and I could hear that he was still carrying wounds even as he counted his reasons to stay sober. I have come to believe that a deep connection with another person is one of the most powerful forces there is for making you finally face yourself. In the last days of March 2016, with only a handful of sober days to my name, his story kept echoing in me. If he could overcome his demons, then so could I.

What followed was not glamorous. I had a job, but I was, for all practical purposes, homeless. For three months I slept on a friend's living room floor, maybe four hours a night, saving every cent I could toward a security deposit and a first month's rent. When I finally got a place of my own and a door I could close, I laid down on the floor and slept for nearly sixteen hours. For a year and a half I had not known, on any given night, where I would sleep the next. I cannot fully put into words what that kind of uncertainty does to a person, in the body and in the mind.

Three months sober, I still felt like I could slip off the edge at any moment. One quiet evening I gathered a few photographs from happier times and set them on my desk. I cannot overstate what those images did for me. A single glance was often enough to pull me back from the edge, to remind me what I was fighting for and who I wanted to become. They were anchors, and I will come back to that practice in a moment, because it matters for you too.

I leaned on Meditations again in that season, reflecting even when I was too raw to journal. I worked a second shift, leaving home around 2:30 in the afternoon and getting back in the small hours. I gave AA a try, and it did not resonate with me, so I stopped. What I could not avoid was the wreckage I had to clean up: three DUIs. My first hearing landed, of all days, on my birthday at the end of August. Over the following months came plea agreements, a stretch in jail during Thanksgiving week, court-ordered counseling, parole, monthly drug and alcohol tests, and grueling early-morning drives to meet my probation officer after late nights at work. It was exhausting in every direction.

By the end of 2016 I understood that I was not going to hold my sobriety together without a system. On December 31st, 2016, right around midnight as 2017 arrived, Flavors of Confidence was born. At the time, my own idea for how to stay sober struck me as a little insane. Time has taught me that the ideas worth keeping often feel that way at first.

In the years that followed, my career climbed back to heights I had not seen in over a decade, and still, something was missing. I was good at whatever I took on, and I was not fulfilled. I was a robot: up in the morning, through the routines, into bed at night, with a hollow place I could not name. I caught myself asking, Is this all there is, sober? If you have ever felt some version of that, you are not alone.

Then, on August 1st, 2020, Samantha was murdered.

I need to tell you who she was, because without her there is no S.O.B.E.R. Method, and there is certainly no book in your hands right now. Samantha was the love of my life. She was the person whose belief in me outran my belief in myself. She was also, years before any of this, the one who sat with me and worked out where the world was heading. Back in 2014 the two of us talked, with a clarity that still unsettles me, about what smartphones and social media were going to do to people: to our attention, our relationships, and our confidence, especially in the young. She is the reason this particular book exists. And for years she had gently, persistently encouraged me to write my story, to put down on paper how I had found my way out by discovering my Flavor of Confidence.

Her death broke something open in me. My sobriety held, but the question of why I was still alive grew louder: survivor's guilt. My father's health was failing at the same time, and in August 2021 I left my job for a new chapter that, for a while, gave me somewhere to put the grief. The void stayed. Late in 2021 I finally began the memoir Samantha had always wanted me to write, and it took fifteen months. In April 2022, my father died, and not long after, my publisher backed out of the project. I paused and questioned everything again. Sober, stable, successful, and still without the fire I was looking for.

Around that time, I needed to figure out what do to with a condo I owned in Lincoln, Nebraska, and started driving up to remodel what years of neglect had left behind. By mid-October I decided to move there, and I circled December 11th, 2022 on the calendar. All the while, I was tracking the agonizing, repeatedly delayed progress of Samantha's murder trial. Every postponement was a fresh wound. But when a final trial date was at last set, something in me caught fire, and I made myself a promise: I would finish the memoir before the trial ended. I did.

We had kept our relationship private for years, which meant her family never knew I had been part of her life or what she had meant to me. I wrote a long, honest letter to her father and sent it with a copy of my book, with no idea what would come back. What came back was warmth. And then her father asked me something I never expected: would I testify at her killer's sentencing? Standing in that courtroom, looking the man who took her in the face, gave me a release I still cannot fully describe. Meeting the people she had told me about felt like meeting family I already knew.

Driving home to Lincoln from Alabama, somewhere just outside Nashville, it hit me like a bolt of lightning. I would spend the rest of my life helping other people find their way to and through their own sobriety, as a tribute to Samantha and everything she had given me. Everything came full circle in that moment. The rest of how the S.O.B.E.R. Method came to be is in the Preface.

I am telling you all of this for a reason beyond honoring her, although I will always take the chance to do that. I am telling you because I have been where you are. Not in your exact struggle, but in the place underneath it: the compulsion you cannot simply will away, the shame, the relapses, the slow and unglamorous work of building a life you do not need to escape from. I did not climb out in one heroic leap. I climbed out by getting a little better, again and again, and by refusing to quit on the days it would have been easier to. That is the whole of it. That is continuous improvement.

Visual Anchors and Their Power

I just told you about the photographs I set on my desk in those early months, and how a single glance could pull me back from the edge. That was not sentimentality. It was evidence of what I was fighting for, proof that the life I wanted was real and not imaginary, and it turns out there is solid neuroscience underneath why it worked.

For digital sobriety, your visual anchors look different. They might be:

» A photograph of someone you love, placed deliberately somewhere you will see it during the times you most tend to reach for your phone, such as by the charger in the kitchen, on the nightstand in place of the phone, or on your desk at work

» A note in your own handwriting, placed where you will see it, with a single sentence from your personal vision statement

» The phone-free zone signs you might post at your dining table or bedroom door

» The old alarm clock on your nightstand that replaced your phone

The power of visual anchors is not sentimental. It is neurological. Visual cues associated with strong positive experiences activate the prefrontal cortex, the seat of deliberate decision-making, and provide a competing stimulus to the automatic pull of the habit. They remind you, in a non-verbal, immediate way, of why you are doing this. They give your willpower something to grab.

Place yours deliberately. Make them visible for your highest-vulnerability moments.

Massimo's S.O.B.E.R. Essentials for Digital Sobriety

These are not new: they are drawn from everything in this book. I am collecting them here because in the daily reality of continuous improvement, a short reference list has value.

1. **Set specific, measurable, date-certain goals.** Not intentions. Goals with dates, metrics, and plans.
2. **Stay connected.** Build and maintain a support network of people who know what you are working on. Tell at least one person. Ask for accountability.
3. **Move your body every day.** Physical presence is the antidote to digital absence. Five days per week minimum.
4. **Journal physically.** Every day. Screen Time data. Progress. Triggers. Gratitude. One thing you experienced today that could not be photographed or shared.
5. **Read Meditations regularly.** One passage per day. Reflect. Write about how it applies to your digital life. Return to the ones that challenge you most.
6. **Maintain your device-free zones.** Bedroom. Dining table. The first hour of your morning. These are not negotiable. They are the foundation.
7. **Review your BCAs.** Weekly at minimum. Update them as your challenges evolve. Add new ones when new trigger patterns emerge.
8. **Celebrate milestones.** Thirty phone-free mornings. Sixty nights with your phone in the kitchen. One hundred days of phone-free dinners. These are real accomplishments. Mark them.
9. **Accept setbacks without shame.** You will slide back. The method is continuous improvement, not perfection. Every return to the cycle is forward motion.
10. **Mentor someone.** When you are ready, when you have six months of genuine progress, find someone who is where you

were and share what you have learned. Helping others is the most reliable way to consolidate your own gains.

11. **Never lose sight of what you are building.** Your Flavor of Confidence in the digital age is not a smaller screen-time number. It is the lived, daily experience of choosing presence over compulsion. Of trusting yourself. Of being actually, genuinely, fully in your own life.

The S.O.B.E.R. Cycle: Always in Motion

A note on the cycling nature of this method, because it is the thing people most commonly misunderstand.

You do not complete the S.O.B.E.R. cycle and arrive at a place of permanent, effortless digital sobriety. You run the cycle continuously, and with each pass, you go deeper, address more subtle challenges, and build more nuanced capacities. The first cycle addresses your most obvious and disruptive device habits. The second cycle addresses the triggers that survived the first. The fifth cycle, two years in, addresses the small ways your relationship with technology is still not fully under your own authority.

This is not a sign of failure. It is the nature of the work, and it is also the nature of human growth. Marcus Aurelius ran his own version of this cycle for decades. His Meditations are the journal of a man who never stopped working on himself, never declared himself finished, and never confused progress with completion. The method he practiced, daily self-examination, honest acknowledgment of failures, recommitment to virtue, repeat, is the oldest continuous improvement cycle in the human record.

You are in good company.

The view of yourself changes with each cycle. The person who returns to Stoic for the third time is not the same person who began it for the first time. You carry your gains with you. The confidence you have built, the presence you have recovered, the relationships you have deepened: these are not erased by a bad week or a slip back into old habits. They are the foundation everything else is built on.

Look at yourself in the mirror and keep growing until you love the person looking back at you.

That is your Flavor of Confidence.

Massimo Rigotti

Notes

NOTES

Notes

Resources and Support

S.O.B.E.R. Method

S.O.B.E.R. Method: https://sobermethod.com

S.O.B.E.R. Method Community: https://sobermethod.org

S.O.B.E.R. Method Podcast: https://thesobermethodpodcast.com

All Socials: @sobermethod

Tag your progress: #sobermethod

Follow Massimo

https://massimorigotti.com

On X @massimorigotti

Deep Thought Friday: On X @dtfriday FB/IG @deepthoughtfriday

Digital Well-Being Tools and Apps

How to enable grayscale:

» **iPhone: Settings** → Accessibility → Display & Text Size → Color Filters → Enable with Grayscale

» **Android: Settings** → Digital Wellbeing → Bedtime Mode → Grayscale or Monochromacy (or search Settings for "Grayscale")

I recommend starting with grayscale during your highest-vulnerability periods: the first thing in the morning, the last hour before bed, during meals. Some people extend it to their entire day.

Built-In Platform Tools:

» iOS Screen Time (Settings → Screen Time): Daily and weekly usage reports, app limits, downtime scheduling, communication limits

» Android Digital Wellbeing (Settings → Digital Wellbeing & Parental Controls): Dashboard, app timers, Focus Mode, Bedtime Mode with grayscale

» Focus Mode (iOS and Android): Silences specified apps for set periods; available through Screen Time (iOS) and Digital Wellbeing (Android)

Helpful Third-Party Apps:

I remain philosophically ambivalent about fighting device addiction with device-based tools. The apps above can be useful as transitional scaffolding, but the long-term goal is self-regulation without app enforcement. Use them as training wheels, not permanent infrastructure.

» **Forest:** Grows a virtual tree while you stay off your phone; trees die if you unlock your device. Available iOS and Android.

» **Flipd:** Locks selected apps for set periods, providing "commitment device" functionality for your phone-free windows.

» **One Sec:** Inserts a brief breathing pause before opening specified apps, interrupting the automatic pickup-and-open habit loop.

» **Freedom:** Blocks distracting apps and websites across all devices simultaneously, useful for the desktop component of device overuse.

» **Opal:** Focus sessions with detailed analytics; app blocking with customizable schedules.

» **ScreenZen:** Inserts mindful pauses before app opening; shows you how many times you have opened an app today.

Books

Essential:

» *Meditations* by Marcus Aurelius (Gregory Hays translation, Modern Library/Penguin): Your primary Stoic text. Not optional. Read one passage per day, every day.

» *The Anxious Generation* by Jonathan Haidt (Penguin Press, 2024): The most thorough and accessible account of how smartphones and social media have reshaped human well-being and what can be done about it.

» *Dopamine Nation: Finding Balance in the Age of Indulgence* by Anna Lembke (Dutton, 2021): Stanford addiction medicine physician's accessible account of the dopamine system and why compulsive digital use shares its architecture with substance addiction.

Strongly Recommend:

» *Flavors of Confidence: A Reflection for Those in Need* by Massimo Rigotti (Rigotti Press): The author's memoir, foundational for understanding the origin of the Flavors of Confidence framework.

» *Digital Minimalism: Choosing a Focused Life in a Noisy World* by Cal Newport (Portfolio/Penguin, 2019): Philosophical and practical case for intentional technology use; complements the S.O.B.E.R. approach.

» *How to Break Up with Your Phone* by Catherine Price (Ten Speed Press, 2019): A practical 30-day program; useful as a companion guide for the Behavior step.

» *Indistractable: How to Control Your Attention and Choose Your Life* by Nir Eyal (BenBella Books, 2019): Practical framework for managing internal and external triggers; written by a former persuasive technology consultant.

» *Stolen Focus: Why You Can't Pay Attention and How to Think Deeply Again* by Johann Hari (Crown, 2022): Comprehensive investigation into the attention crisis, with extensive research on the tech industry's role.

Support Communities and Organizations

Digital Wellness and Screen-Time Support:

» Center for Humane Technology: https://www.humanetech.com, Resources on technology design and well-being; produced the documentary *The Social Dilemma*

» Common Sense Media: https://www.commonsensemedia.org, Research and resources on technology use for families; excellent for parents

» Digital Wellness Institute: https://digitalwellnessinstitute.com, Professional certification and research hub

For Parents:

» Wait Until 8th: https://www.waituntil8th.org, Campaign for delaying smartphones until 8th grade

» Screenagers: https://www.screenagersmovie.com, Documentary and resources for families navigating tech and adolescents

Mental Health Support:

» Anxiety & Depression Assoc. of America: https://adaa.org

» Crisis Text Line: Text HOME to 741741 (U.S.)

» SAMHSA National Helpline (U.S.): 1-800-662-4357 (for co-occurring substance and behavioral addictions)

Crisis Hotline:

» United States – 988 Suicide & Crisis Lifeline: Call or text 988 or 988lifeline.org

» United Kingdom – Samaritans: Call 116 123 free from any phone

» Australia – Lifeline: Call 13 11 14

If you or someone you know is in crisis or having thoughts of suicide. **Please reach out for help.** A good starting point with localized resources for many countries is: https://www.iasp.info/suicidalthoughts/

If you are in immediate danger, contact your local emergency services.

References

1. Reviews.org. (2023). *Cell Phone Usage Statistics 2023: Mornings Are for Notifications.* Reviews.org.
2. Lam, S.S.M., et al. (2024). "Problematic smartphone use and psychological well-being: A comprehensive meta-analysis of 94 studies." *Journal of Behavioral Addictions*, 13(1), 45–68.
3. Scott, H., et al. (2022). "Nighttime smartphone use and sleep quality: A meta-analysis of 125,000 participants." *Sleep Medicine Reviews*, 63, 101611.
4. Haidt, J. (2024). *The Anxious Generation: How the Great Rewiring of Childhood Is Causing an Epidemic of Mental Illness.* Penguin Press.
5. Mueller, P.A., & Oppenheimer, D.M. (2014). "The pen is mightier than the keyboard: Advantages of longhand over laptop note taking." *Psychological Science*, 25(6), 1159–1168.
6. Andrews, S., et al. (2015). "Beyond self-report: Tools to compare estimated and real-world smartphone use." *PLOS ONE*, 10(10), e0139004.
7. Data.ai (formerly App Annie). (2024). *State of Mobile 2024.* Data.ai.
8. Fogg, B.J. (2019). *Tiny Habits: The Small Changes That Change Everything.* Houghton Mifflin Harcourt.
9. Roberts, J.A., & David, M.E. (2023). "Partner phubbing and relationship quality: A longitudinal study." *Psychology of Popular Media*, 12(3), 310–324.
10. Yildiz, M.A. (2022). "Parental smartphone use and child attachment security: Mechanisms and outcomes." *Current Psychology*, 41, 7811–7823.
11. Greenfield, D.N. (2015). *Treatment Considerations in Internet*

and Video Game Addiction: A Qualitative Discussion. Center for Internet and Technology Addiction.

12. Chang, A.M., et al. (2015). "Evening use of light-emitting eReaders negatively affects sleep, circadian timing, and next-morning alertness." *PNAS*, 112(4), 1232–1237.
13. Walker, M. (2017). *Why We Sleep: Unlocking the Power of Sleep and Dreams.* Scribner.
14. Exelmans, L., & Van den Bulck, J. (2024). "Bedtime mobile phone use and sleep outcomes: A large-scale study of 14,000 adults." *Sleep Medicine*, 115, 44–52.
15. Mark, G. (2023). *Attention Span: A Groundbreaking Way to Restore Balance, Happiness and Productivity.* Hanover Square Press.
16. Drouin, M., et al. (2012). "Phantom vibrations among undergraduates: Prevalence and associated psychological characteristics." *Computers in Human Behavior*, 28(4), 1490–1496.
17. Dunbar, R.I.M. (2018). "Do online social media cut through the constraints that limit the size of offline social networks?" *Royal Society Open Science*, 5(1), 150292.
18. Verduyn, P., et al. (2015). "Passive Facebook usage undermines affective well-being: Experimental and longitudinal evidence." *Journal of Experimental Psychology: General*, 144(2), 480–488.
19. Primack, B.A., et al. (2017). "Social media use and perceived social isolation among young adults in the U.S." *American Journal of Preventive Medicine*, 53(1), 1–8.
20. Pieh, C., et al. (2025). "Smartphone screen time reduction improves mental health: a randomized controlled trial." *BMC Medicine*, 23, 107.
21. Kushlev, K., & Dunn, E.W. (2015, replicated 2019). "Checking email less frequently reduces stress." *Computers in Human*

Behavior, 43, 220–228.

22. Andrews, C., et al. (2024). "Grayscale phone screen settings and smartphone use: A randomized controlled trial." *PLOS ONE*, 19(2), e0298764.

23. Ward, A.F., et al. (2017). "Brain drain: The mere presence of one's own smartphone reduces available cognitive capacity." *Journal of the Association for Consumer Research*, 2(2), 140–154.

24. Emmons, R.A., & McCullough, M.E. (2003). "Counting blessings versus burdens: An experimental investigation of gratitude and subjective well-being in daily life." *Journal of Personality and Social Psychology*, 84(2), 377–389.

25. American Academy of Sleep Medicine. (2014). *Healthy Sleep Habits*. AASM Foundation.

26. Jenkinson, C.E., et al. (2013). "Is volunteering a public health intervention? A systematic review and meta-analysis of the health and survival of volunteers." *BMC Public Health*, 13, 773.

27. Post, S.G. (2005). "Altruism, happiness, and health: It's good to be good." *International Journal of Behavioral Medicine*, 12(2), 66–77.

28. Ferster, C.B., & Skinner, B.F. (1957). *Schedules of Reinforcement*. Appleton-Century-Crofts.

29. Lembke, A. (2021). *Dopamine Nation*. Dutton.

30. Gallup & Walton Family Foundation. (2024). *Voices of Gen Z: Perspectives on U.S. Education, Well-Being, and the Future (Year 2 Report)*. Gallup.

31. Twenge, J.M., et al. (2022). "Worldwide increases in adolescent loneliness." *Journal of Adolescence*, 93, 257–269.

32. Vahedi, Z., & Saiphoo, A. (2024). "Social media use and psychological well-being: A systematic review of passive versus active use effects." *Computers in Human Behavior*, 152, 108–124.

33. James, W. (1890). *The Principles of Psychology* (Vol. 1, Chapter 11, "Attention"). Henry Holt and Company.

34. Kim, J., et al. (2024). "Color reduction and compulsive smartphone use: Mechanisms and outcomes." *Cyberpsychology, Behavior, and Social Networking*, 27(4), 201–215.

35. Stieger, S., & Lewetz, D. (2024). "Benefits of digital detox interventions: A systematic review." *Cyberpsychology, Behavior, and Social Networking*, 27(3), 145–162.

36. Van Deursen, A.J.A.M., et al. (2025). "Behavioral self-regulation for problematic smartphone use: A systematic review of 31 studies." *Internet Interventions*, 39, 100754.

www.ingramcontent.com/pod-product-compliance
Lightning Source LLC
LaVergne TN
LVHW090409210926
848480LV00002B/504